Universitext

Universitext

Universitext is a series of textbooks that presents material from a wide variety of mathematical disciplines at master's level and beyond. The books, often well class-tested by their author, may have an informal, personal, even experimental approach to their subject matter. Some of the most successful and established books in the series have evolved through several editions, always following the evolution of teaching curricula, into very polished texts.

Thus as research topics trickle down into graduate-level teaching, first textbooks written for new, cutting-edge courses may make their way into *Universitext*.

For further volumes:
http://www.springer.com/series/223

Jürgen Jost

Mathematical Methods in Biology and Neurobiology

 Springer

Jürgen Jost
Max Planck Institute for Mathematics
  in the Sciences
Leipzig
Germany

ISSN 0172-5939   ISSN 2191-6675   (electronic)
ISBN 978-1-4471-6352-7   ISBN 978-1-4471-6353-4   (eBook)
DOI 10.1007/978-1-4471-6353-4
Springer London Heidelberg New York Dordrecht

Library of Congress Control Number: 2014931395

Mathematics Subject Classification: 05Axx, 05Cxx, 05C05, 05C50, 05C75, 05C80, 05C82, 31B05, 34A12, 34A34, 34C05, 34C11, 34C15, 34C23, 34C25, 34C26, 34C28, 34C29, 34C60, 34D20, 34D23, 34D45, 34E10, 34E13, 35A01, 35B36, 35B50, 35B51, 35C07, 35G05, 35G10, 35G15, 35J05, 35K57, 35K05, 35L05, 35Q83, 35Q92, 49J05, 60G07, 60G55, 60J60, 60J70, 60J80, 60J85, 70K70, 92B05, 92C05, 92C40, 92C42, 92D15, 92D25, 92E20.

Mathematica® is the registered trademark of Wolfram Research, Inc., http://www.wolfram.com/

Printed on acid-free paper

Springer is part of Springer Science+Business Media (www.springer.com)

Preface

Structures and processes studied in biology range from molecules in cells to populations or ecosystems, or to brains consisting of billions of interacting neurons, and the formal models employed in biology range from graphs as abstract representations of pairwise interactions to complicated systems of partial differential equations that try to capture all details of some biological system. Therefore, also the mathematical methods and tools employed in biology and neurobiology are quite diverse and heterogeneous. A student wanting to learn and apply mathematical techniques in biology might be confronted with the problem that she or he does not possess an overview of the available mathematical tools and does not know which method could be appropriate for a specific biological problem. A biological structure, in fact, can be modeled at various levels of details, and it is not necessarily the case that a more detailed and precise model yields better quantitative predictions.

In that situation, this book presents a spectrum of mathematical methods that are relevant and important for biology and neurobiology. Thereby, the student should be equipped with an overview and a working knowledge of the most important mathematical tools. These methods fall into three categories: First of all, there are the discrete methods, from combinatorics and graph theory. Graphs can be used to model the structure of pairwise interactions between elements in some network, whatever their precise biological nature might be. They can also be utilized to analyze empirical network data. A particular class of graphs, the trees, plays a special role in biology because they model descendence relations. The second class of models comprises the stochastic ones. Much of biology, in fact, is modeled in stochastic terms, be it the firing of neurons in the brain or the random forces of evolution. Therefore, I provide a systematic introduction to stochastic processes. Finally, there are the analytical methods from the theory of differential equations, like dynamical systems or partial differential equations, that are used to explain the formation of biological patterns, ranging from the molecular scale to that of interacting species. Often, such models are derived from optimization principles, the theoretical rationale being that evolutionary competition has produced structures that best perform certain functions. Therefore, we also devote a chapter to optimization schemes. A final chapter then deals with a particular area of mathematical biology, population genetics. That has been the field of biology where mathematical methods first have been applied in a very systematic manner. It continues to be alive today, and I present a new geometric approach to population genetics that will, as I hope, clarify the underlying mathematical structure.[1] These last two chapters are thus both concerned with issues of evolutionary biology, but from two different perspectives, that of optimization versus that of random processes. For a mathematical understanding of evolution, the combination of these two perspectives is essential.

The exercises are concerned with both the mathematical techniques developed in this book and their application in biological modeling. While some exercises are more of the traditional drill type that is needed to master some technique, others are more open in order to stimulate and encourage your own thinking.

In this book, I try to explain the underlying mathematical concepts and to prove the easier statements so that the reader can develop some feeling for the abstract mathematical structures. Throughout the text, I also develop applications to biology, from intracellular structures or the dynamics of neurons to those of populations. Thus, the applications span many physical orders of magnitude, but perhaps somewhat surprisingly, often the same mathematical structures turn out to yield useful models at several rather different levels. In any case, the systematic arrangement of the material is according to mathematical and not to biological principles. This seemed the natural choice for the material to be presented here, but in order to compensate for that, I am writing a companion volume "Biology and Mathematics" [2] where I attempt a systematic presentation according to biological principles and structures. Actually, I have recently also written a book entitled "Mathematical Concepts" [3] where the mathematical structures are developed at a much more abstract level. I believe that this may be relevant for biology because theoretical biology needs to develop more abstract and encompassing concepts in order to organize and understand the multitude of biological structures and processes and the increasing wealth and heterogeneity of biological data more deeply. This aspect, however, is not addressed in the present book which rather concentrates on established mathematical methods and their biological applications. In contrast to such an abstract systematic treatment, this book emphasizes the richness and diversity of the applications of mathematics to biology.

The literature in mathematical biology is too extensive to be adequately covered in this book. Therefore, the references are very selective, and you should consult the monographs and survey articles listed in the bibliography for further or more precise references. I apologize to any authors whose work is not, or not correctly, referenced in this book.

In any case, while this book certainly aims at teaching a range of mathematical concepts and methods that are relevant for the modeling and analysis of biological structures and processes, it also wants to stimulate your curiosity about biological phenomena and your independent thinking about how to model and analyze them with mathematical tools.

---

[1] A more detailed exposition of this theory will be given in [1].

This book is based on graduate courses at Leipzig (in a joint program between the Max Planck Institute for Mathematics in the Sciences and the Department of Mathematics and Computer Science of Leipzig University, the International Max Planck Research School "Mathematics in the Sciences," directed by Stephan Luckhaus) and at the Ecole Normale Superieure in Paris (organized by Benoît Perthame).

Thus, a student who would like to use this book should have some basic mathematical knowledge, including in particular calculus. Some background in biology might help to appreciate the significance of the mathematical methods, but is not indispensable for reading this book. In fact, the book can also be taken as a survey over a rather wide range of mathematical structures, for any student of mathematics or the sciences.

I thank Ugur Abdullah, Fatihcan Atay, Nihat Ay, Anirban Banerjee, Frank Bauer, Nils Bertschinger, Pierre-Yves Bourguignon, Olaf Breidbach, Andreas Dress, Bernhard Englitz, Boris Gutkin, Julian Hofrichter, Danijela Horak, Bobo Hua, Martin Kell, François Kèpés, Ilona Kosziuk, Michael Lachmann, Shiping Liu, Stephan Luckhaus, Vic Norris, Eckehard Olbrich, John Pepper, Benoît Perthame, Johannes Rauh, Christian Rodrigues, Thimo Rohlf, Areejit Samal, Klaus Scherrer, Susanne Schindler, Peter Schuster, Bärbel Stadler, Peter Stadler, Angela Stevens, Tat Dat Tran, Henry Tuckwell, Leo van Hemmen, and others whom unfortunately I may have forgotten to mention, for various discussions about the mathematical and/or biological aspects. Tat Dat Tran also checked the entire manuscript, pointed out several corrections, and created some of the figures, in particular the simulations of the FitzHugh-Nagumo and van der Pol equations, with the help of Mathematica®. For several of the diagrams, I have used the latex supplement DCpic of Pedro Quaresma.

The author also gratefully acknowledges the support from the ERC Advanced Grant FP7-267087, the Volkswagenstiftung and the Klaus Tschira Stiftung.

References

1. Hofrichter, J., Jost, J., Tran, T.D.: Information geometry and population genetics, in preparation
2. Jost, J.: Biologie und Mathematik, to appear
3. Jost, J.: Mathematical Concepts, to appear

Contents

Chapter 1
Introduction

Abstract
Questions:

- What can mathematics contribute to biology, and which mathematical theories are useful for that purpose?

Biology does not have the clear structure of mathematics. Nevertheless, it possesses some fundamental concepts. The *gene* is the unit of coding, function, and inheritance. It contains the information for a phenotypic trait that is realized in interaction with contributions from the environment and transmitted to offspring. The *cell* is the basic unit within which metabolic processes can take place. The *species* is the dynamic pool for genetic recombination. An *organism* is a carrier of genes, an organized ensemble of cells and a member of a population or species. Mathematical methods to study biological phenomena can be taken from algebra, analysis, stochastics, or geometry, but should always be developed with a clear vision of the biological problems to be addressed.

1.1 Theses About Biology

Thesis 1. *Biological structures are aggregate structures. Therefore, biological laws are not basic ones that do not admit exceptions, but rather emerging from some lower scale.*

Thesis 2. *Biological entities are discrete, but biological structures are situated in continuous space and biological processes take place in continuous time.*

Thesis 3. *Biological processes intertwine stochastic effects and deterministic dynamics. Randomness can support order while deterministic processes can be unpredictable, chaotic. The question then is at which level regularities emerge.*

J. Jost, *Mathematical Methods in Biology and Neurobiology*, Universitext,
DOI: 10.1007/978-1-4471-6353-4_1, © Springer-Verlag London 2014

Thesis 4. *Large populations of discrete units can be described by continuous models and, conversely, invariant discrete quantities can emerge from an underlying continuous substrate.*

Thesis 5. *Fundamental biological concepts, like fitness or information, are relative and not absolute ones.*

Thesis 6. *Fundamental biological quantities do not satisfy conservation laws. Those rather appear as external constraints.*

Thesis 7. *Biological systems interact with their environments and are thermodynamically open. Biological structures sustain the processes that reproduce them and are therefore operationally closed.*

Thesis 8. *Biological structures are results of historical processes. It is the task of biological theory to distinguish the regularities from the contingencies.*

Thesis 9. *The abstract question posed to mathematics by biology is structure formation. This needs to be understood as a process because living structures are not at thermodynamic equilibrium.*

Thesis 10. *Gathering biological data without guiding concepts and theories is useless.*

1.2 Fundamental Biological Concepts

1. The **gene** is the unit of coding, function, and inheritance. As such, it links molecular biology and evolutionary biology. The Neodarwinian Synthesis combined Mendel and Darwin. Modern molecular biology seems to offer a more basic perspective.
2. The **cell** is the unit of metabolism. It constitutes the basic operationally closed, autopoietic system in biology. Modern biology struggles to understand cells on the basis of their molecular constituents, DNA, RNA, and polypeptides (proteins). Multicellular organisms emerge through a partial suppression of the autonomy of the constituting cells.
3. The **species** represents the balance between the diverging effects of genetic mutations and selection at the organismic or other levels and the converging mechanism of sexual recombination. It is the arena of population biology, a child of the Neodarwinian Synthesis and the first success of mathematical models in biology. It is also important in ecology.

The **organism**, in fact, is the carrier of genes, the organization of cells, and the member of a species. It thus links the three fundamental biological concepts. It is also a, but not the exclusive, unit of selection.

It seems that neurobiology has not yet identified such a fundamental concept, but perhaps the **spike** can be considered as the basic event of information transmission, and the **synapse** as the basic structure supporting this.

1.3 A Classification of Mathematical Methods

The following is a somewhat incomplete list, arranged partly with relevance for biology in mind.

1. Discrete structures $\to$ **Algebra**
   (a) Static structures
       i. Algebraic concepts: Combination and composition of objects
       ii. Graphs and networks, including phylogenetic trees
       iii. Information
       iv. Discrete invariants of continuous structures and dynamical processes
   (b) Discrete processes (Cellular automata, Boolean networks, finite state machines,...)
   (c) Game theory as the formalization of competition

2. Spatial relations $\to$ **Geometry**
   (a) Geometry of (three-dimensional) physical space
   (b) Abstract notions of space for expressing relationships (discrete ones like graphs and continuous ones like Hilbert spaces; state spaces of dynamical systems)
   (c) Symmetries and invariances

3. Continuous methods $\to$ **Analysis**
   (a) Deterministic dynamical processes
       i. Continuous states enable phase transitions and bifurcations, that is, qualitative structural changes resulting from small underlying variations
       ii. Continuous states and time: Ordinary differential equations and other dynamical systems
       iii. Continuous spatial structures: Partial differential equations (example: Reaction-diffusion equations)
   (b) Stochastic analysis
       i. Stochastic processes (while stochastic processes may also operate on discrete quantities, the concept of probability is a continuous one)
       ii. Population processes: averaging over stochastic fluctuations in lower level dynamics
       iii. Optimization schemes with stochastic ingredients: Genetic and other evolutionary algorithms, swarm algorithms for distributed search, certain neural networks,...
       iv. Statistical methods for the analysis of biological data

4. Hybrid models
    (a) Difference equations (continuous states, but discrete time)
    (b) Dynamical networks (dynamical systems coupled by a graph), in particular
         neural networks

5. System theory as a global unifying perspective?

According to the preceding list, not all mathematical subjects seem to be relevant for biology. Classical algebraic structures occur in a cursory manner at best, and one of the deepest branches, number theory and arithmetics, is entirely absent. Three-dimensional physical space constitutes an important constraint for biological organization. Organisms and their constitutive biological structures like cells are living and interacting in space, and are defining and shaping their own spaces like architectural structures which is constitutive for morphology. Symmetries and invariances, the merging ground of algebra and geometry, are important issues for the neurobiology underlying cognition, as well as for many classification purposes. In any case, the branches of algebra, geometry and analysis are often interwoven.

Chapter 2
Discrete Structures

Abstract
Questions:

- How can the cells of an organism which all share the same genes can fulfill so many different functions?
- Are there good mathematical tools to identify the important features in all those networks that modern biological data collection produces?
- How long ago did the last common ancestor of two species or two individuals live?

A model of combinatorial gene regulation shows the power of combinatorics. Graphs are useful tools for network analysis, and their spectral theory is developed. Phylogenetic relationships between species are modeled by particular types of graphs, the trees. Descendence relations between individuals involve two parents and lead to genealogies. Coalescents treat the question of common ancestors. Such structures also naturally lead to the stochastic processes treated in the Chap. 3.

2.1 Introductory Example: Gene Regulation and the Power of Combinatorics

In this section, I present an example of a combinatorial scheme in molecular biology. This is meant to show that even elementary mathematical reasoning can help us to clearly understand a biological situation that may initially look rather complicated. First, however, I shall sketch the most basic principles of molecular biology. More details can be found in standard textbooks, like [1] or [93].

Metabolism and other fundamental functions of the cell are essentially carried out by proteins. The building blocks of proteins are polypeptides, sequences of typically a few hundred amino acids that fold into particular three-dimensional shapes according to attractive forces between different amino acids and interactions with water molecules in the cell. A protein consists of one or several such polypeptides, and

J. Jost, *Mathematical Methods in Biology and Neurobiology*, Universitext,
DOI: 10.1007/978-1-4471-6353-4_2, © Springer-Verlag London 2014

its three-dimensional shape determines its function. The information for the particular sequence of each polypeptide is contained in the DNA of the cell. The DNA is a sequence itself, consisting of nucleotides instead of amino acids, and the DNA is inherited by the daughter cells under cell division and the germ cells in sexual recombination. This will now be described with some more details and precision.

The fundamental process of molecular biology then is gene expression, that is, the production of polypeptides, the building blocks of proteins, according to the genetic information contained in the DNA of a cell. The DNA (desoxyribonucleic acid) is a long string of base pairs, arranged in the shape of a double helix, as discovered by Watson and Crick. There are four different nucleotide bases, labelled A, C, G, and T (we are not concerned here with their precise chemical identity, and so, these letters may suffice for our purposes). Thus, each of the two strands of the double helix is a long sequence composed of these 4 "letters". Each strand determines the identity of the complementary strand, because C in one strand is paired with G in the other, and A with T. Therefore, when the double helix is split apart, each strand contains the complete information for assembling a new such double helix. This is the principle underlying genetic inheritance. Here, however, we are not concerned with inheritance, but rather with gene expression. The first step of gene expression, called transcription, then consists in copying the information in a segment of one of the strands into another macromolecule, RNA (ribonucleic acid), which is chemically more active and flexible. It also consists of sequences composed of 4 letters, A, C, G as in the DNA and a new letter U taking the place of T. Again, this copying works according to the above complementarity principle. Which segments of the DNA are thus copied under a given cellular condition is controlled by certain proteins, the transcription factors that typically bind to locations in the DNA nearby those to be copied and that can then trigger, enhance or block the transcription process [26]. Of course, one and the same stretch of DNA can be repeatedly transcribed, and the regulation of the number of such transcripts is essential, but we shall not emphasize this aspect in the sequel. The resulting RNA is then further processed, through interactions with itself or with other RNAs or with certain proteins again. The final mRNA (m standing for "messenger") can then be translated into a polypeptide, in a certain complex called the ribosome, with the help of some other auxiliary RNA, the rRNA (r standing for "ribosomal"). The principle of the translation is that the unit of translation in the mRNA is a triplet of nucleotides, like ACG or UAA, also called a codon. Each such triplet is translated into a specific amino acid, and the resulting polypeptide thus is a sequence of amino acids. Since there are 64 possible triplets, but only 20 amino acids, several different triplets can correspond to the same amino acid. This fact is called the degeneracy of the genetic code, although redundancy might be the more accurate word. (Actually, the triplet UGA has a special role: It serves as the stop codon, that is, when this triplet is encountered in the ribosomal complex, the polypeptide is released, and a new translation can start.) In fact, the relation between such triplets and amino acids is mediated by another type of RNA, called tRNA (t for "transfer"). Chemically, this relation, called the genetic code, that is, which triplet is translated into which amino acid, is arbitrary, and so the question emerges why the translation rules are as they are, instead of being different. That is, why is for instance

GCC translated into the amino acid alanine, instead of, say, cysteine? Is that simply a historical accident, an arbitrary rule that all living creatures have inherited from their common ancestor who had adopted these translation rules by chance? Or are there some chemical or formal principles behind this, like symmetry considerations or coding efficiency? There have been many different speculations about this issue, but none so far has met with general approval.

One or more polypeptides then are combined into a protein. An important point is that a protein is not simply an amino acid sequence, but that for its molecular function, it assumes a specific three-dimensional shape. This shape, is determined by chemical attraction and repulsion between different pieces, but the details are very intricate, and the problem of computing the three-dimensional shape of a protein, or better, the process, called protein folding, by which it acquires this shape from its constituting amino acid sequence is not yet fully solved, despite considerable attempts by many mathematicians and physicists.

The fundamental question for a cell then is which genes to express when, under which circumstances. The mechanism of the cell for answering this question is gene regulation. I have already described that specific proteins, the transcription factors, trigger or inhibit the transcription of DNA segments. In eukaryotic cells (the cells that we are made of, those containing a nucleus, in contrast to prokaryotic cells, without nucleus, like bacteria), the most important part of gene regulation, however, seems to take place at the level of RNA rather than DNA. First of all, the transcribed RNA, called pre-mRNA, is reassembled in a process called splicing into mRNA. Here, on one hand, certain segments, the so-called introns, are cut out whereas the remaining ones, the exons, can then be assembled possibly in different ways, so as to produce different results from one and the same stretch of DNA [13], or pieces of different origins can be put together or interact in other ways. The processing on one hand is based on the spatial configuration assumed by an RNA molecule, on the basis of bindings between complementary nucleotides (A with U or C with G), no longer between different strands as in the DNA, but now between bases in one and the same RNA sequence [59]. On the other hand, it results from interactions with certain other small RNAs, the so-called miRNAs (mi for "micro") or siRNAs (si for "small interfering" or "silencing") or with specific proteins. These proteins bind to RNA molecules to form so-called RNP complexes (where P stands for "protein") [119]. Much of this RNA regulation works as repression, that is, preventing the mRNA from being translated. The biological rationale for this is that on one hand, the production of RNA is energetically cheap, and on the other hand, with mRNA already around, it is much faster to produce the corresponding proteins than if the process had to start anew from the DNA level. Thus, the cell can respond much quicker to new circumstances. (For a systematic analysis, see [103, 104] and the subsequent discussion in the journal Theory in Biosciences, see [105].)

After the genome of humans (and several other species) has been sequenced, that is, the the identity of all the 3 billion letters in the DNA sequence has been established [63], now the ENCODE project systematically records and catalogues all the different RNA molecules that can be present in human (and other) cells [34, 38, 49]. The genetic sequence contains both coding information that can be potentially activated and utilized in a cell with the assistance of specific proteins, and important structural elements. But we need to identify all the different RNAs and understand their interactions with other RNAs and proteins in order to understand the regulation of gene expression in the active cell.

Now, obviously, the scheme described offers many possibilities for combinatorial reasoning as a formal description of the rules governing those processes. Here, as an example I shall discuss a model that arises from my work with the molecular biologist Klaus Scherrer, see [76]. The important point here is that the nucleotides in an mRNA can assume two different roles simultaneously. On one hand, they are parts of coding triplets (except for certain portions at the beginning or end of an mRNA sequence). On the other hand, stretches of about 30 nucleotides can function as binding sites for specific proteins which then regulate the fate of the mRNA, as explained (see [103, 104]). We call such a regulatory stretch of nucleotides an oligomotif. In the basic version of the model, there then is a one-to-one correspondence between such oligomotives and mRNA binding proteins. That is, there is a second, regulatory, code superimposed upon the first code, the genetic code governing translation. In both cases, however, the chemical identity of the nucleotides involved is crucial. An average mRNA may then possess about 20 such oligomotives. The ground state then is when the corresponding proteins are attached to all those 20 oligomotives. In this state, the mRNA is repressed and not translated. It only becomes available for translation when at least 3 of those proteins are removed. (We shall call such a set of 3 oligomotives, or equivalently, of 3 mRNA binding proteins, a triple, not to be confused with the triplet of the genetic code.) That is, when a signal arrives in the cell that causes the release of 3 such binding proteins, the corresponding mRNA gets translated, and a specific polypeptide is produced. Now, however, in a given situation, a cell needs not only one type of polypeptide, but a suitable combination of perhaps hundreds of polypeptides. The preceding structure now offers an elegant scheme for the coordinated expression of groups of genes, that is, the coordinated production of specific combinations of polypeptides and proteins. First of all, there are then $\binom{20}{3} = 1,140$ different possibilities for such triples of oligomotives. The key point now is that different mRNAs will share some, but not all of their oligomotives. That is, whenever we identify 3 proteins for removal, that is, select 3 oligomotives, we then get a specific set of mRNAs that contain those 3 oligomotives and that will then get translated, whereas the remaining ones will stay repressed. And when we select a different set of 3 oligomotives, we obtain a different combination of mRNAs to be translated, hence a different combination of proteins in the cell. This set may partially overlap with the preceding one, depending on the distribution of oligomotives across the different RNAs. In fact, one estimates that there are about 3,000 different mRNA binding proteins, hence also about 3,000 different oligomotives according to the model. We thus have $\binom{3,000}{20}$ different possibilities to distribute the oligomotives across the mRNAs (there are perhaps around 10,000 different mRNAs in a typical mammalian cell).

Let us now look into this scheme in more numerical detail. As explained, in order that several mRNAs participate in the same condition, they need to share at least 3 oligomotives. And when some mRNAs share m oligomotives ($3 \leq m \leq 20$), they can simultaneously participate in $\binom{m}{3}$ conditions. This number varies from 1 (for $m = 3$) to 1,140 (for $m = 20$). However, when $m = 20$, that is, when the mRNAs share all their oligomotives, they can no longer be distinguished in this scheme. Let us consider some numerical examples, on the basis of the general scheme. For K oligomotives, there are $\binom{K}{20}$ different possibilities to choose 20 among them. This means that we can distinguish that many mRNAs through their different endowments with 20 out of these K oligomotives. As explained, a condition for translation is achieved by the selection of 3 (or more) out of these K oligomotives. Every choice of $\binom{K}{3}$ yields a different condition. Precisely those mRNAs will participate in such a condition that carry all those 3 oligomotives. Thus, 3 out of their 20 oligomotives are fixed, and 17 remain for free choice. That is, we have $\binom{K-3}{17}$ different possibilities. Thus, assuming that all the above $\binom{K}{20}$ possibilities are realized, by selecting 3 oligomotives, we select $\binom{K-3}{17}$ different mRNAs. Here are simple numerical examples.

- Distribute 21 oligomotives among 21 mRNAs (20 oligomotives/mRNA) so that each mRNA is identified by which oligo it does not contain. By specifying 3 oligomotives, any of the possible $\binom{21}{3} = \binom{21}{18} = 1,330$ combinations of 18 mRNAs can then be selected. Here, we have only relatively few different mRNAs.
- Distribute 23 oligomotives among $\binom{23}{3} = 1,771$ mRNAs (20 oligomotives/mRNA) so that each mRNA is identified by which 3 oligomotives it does not contain. By specifying 3 oligomotives, any of the possible $\binom{23}{3} = \binom{23}{20} = 1,771$ combinations of $\binom{20}{3} = 1,140$ mRNAS can be selected. Here, we obtain a large collection of selected mRNAs.
- Distribute 22 oligomotives among $\binom{22}{2} = 231$ mRNAs (20 oligomotives/mRNA) so that each mRNA is identified by which 2 oligomotives it does not contain. By specifying 3 oligomotives, any of the possible $\binom{22}{3} = \binom{22}{19} = 1,440$ combinations of $\binom{19}{2} = 171$ mRNAs can be selected. This is a biologically reasonable number.

Obviously, the number 3,000 of different mRNA binding proteins, that is, of different oligomotives is far larger than needed in our model. This indicates that, in reality, gene regulation at mRNA level is more complex than captured by the model. Nevertheless, the model should describe a core principle of regulation. Moreover, there is an interesting combinatorial problem suggested by this model: How to distribute K labels among N units so that each unit receives k of them so that by selecting $\kappa < k$ of them (for which we have $\binom{k}{\kappa}$ different possibilities), we identify the maximal number of different subsets of those N units? We may here wish to constrain those subsets to be of some fixed size n, or to be within a certain size range, say between n_1 and n_2.

In order to understand the mathematical structure of this problem better, it is helpful to translate it into a combinatorial design problem. We consider an $N \times K$ matrix with entries 1 or 0 where each of the N rows has precisely k 1s, and hence $K - k$ 0s. For $\kappa < k$, we then want to find collections of rows that have (at least) κ 1s in common. The question then is how to distribute the 1s in the rows so as to find as many such collections as possible within a given size range.

No full solution seems to be known for this problem. In any case, the example is meant to show that by elementary mathematical reasoning, we can come up with clever ways of how a cell could regulate its genes so that in one situation, in a single stroke, it can co-activate specific groups of genes, and in another situation, again in a single stroke, it can activate another set of genes, perhaps partly overlapping with the first one, without having to address all these genes individually. This is the power of combinatorics.

2.2 Graphs and Networks

2.2.1 Graphs in Biology

A graph is the mathematical structure representing binary relationships between discrete elements. These elements are the vertices of the graph, and the relationships are encoded as connections or edges between vertices. Such a graph can then be a network, that is, the substrate of dynamical interactions carried by the edges between processes located at the vertices. Biological applications abound.

In neural networks, the vertices stand for neurons, and the edges for synaptic connections between them. The interaction is the electrochemical transmission of pulsed dynamical activity, the spikes generated in the neurons. This activity is considered to be the carrier of information, enabling cognitive processes, but the precise identification of the information inside that dynamical activity remains unclear at present. At smaller scales, the vertices can represent molecules like proteins, and the edges again interactions between them. The vertices can also stand for genes, and the edges for correlations in expression patterns indicating functional interactions.

At larger scales, the vertices can be the members of a population, and the edges social or other interactions, like mating. For a population with separate sexes, we then have a bipartite graph, that is, one with two distinct classes of elements such that edges exist only between members of opposite classes, but not inside one class.

At the still larger scale of ecosystems, the vertices can represent species, and the edges stand for trophic interactions. The graph then encodes a food web.

Another important class of biological graphs are the phylogenetic trees that turn genetic or other similarities between species into descendence relations from common ancestors. For individual descendence relations inside a sexually recombining species we rather have pedigrees because each individual then has two parents which in turn may have more than one offspring.

For detailed studies of biological networks and their properties, the reader can consult [94] and [111] and the many references therein.

2.2.2 Definitions and Qualitative Properties

We now display some formal definitions and start with the simplest situation. A graph Γ is a pair (V, E) of a finite set V of vertices or nodes and a set E of unordered pairs, called edges or links, of different elements of V (and we assume $E \neq \emptyset$ to make the graph nontrivial). Thus, when there is an edge $e = (i, j)$ for $i, j \in V$, we say that i and j are connected by the edge e and that they are neighbors, $i \sim j$. Defining edges as unordered pairs of vertices means that we consider (i, j) and (j, i) as the same pair. Thus, the neighborhood relation is symmetric. Requiring that the vertices connected by an edge be different then means that there are no edges connecting a vertex to itself. Thus, the neighborhood relation is not reflexive. In general, it is not transitive either, that is, $i \sim j$ and $j \sim k$ need not imply $i \sim k$. The degree n_i of the vertex i is the number of its neighbors. Also, the order $|\Gamma|$ is the number of vertices in Γ, i.e., the cardinality of the vertex set V.

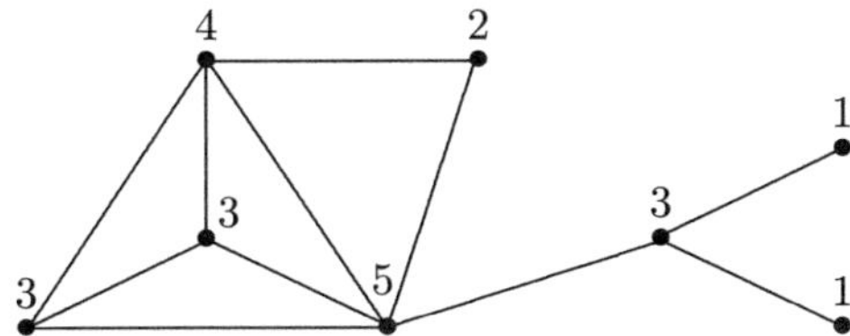

A graph Γ of order 8, with vertex degrees indicated   (2.2.1)

So far, we are assuming that the edges are undirected, that is, the edge (i, j) is the same as (j, i). One may, naturally, also consider directed graphs, that is, where an edge $e = (i, j)$ is considered to go from i to j rather than connect i and j in a symmetric manner. For example, this is appropriate for formalize neurobiological networks because synapses between neurons are directed, starting at the presynaptic neuron and going to the postsynaptic one. In addition, synapses have strengths or weights, and so, we can also consider weighted graphs where each edge e carries a weight or label w_e that indicates its strength. In fact, we may then also allow that some of the weights are negative. In a neural network, an edge with a negative weight would represent an inhibitory synapse.

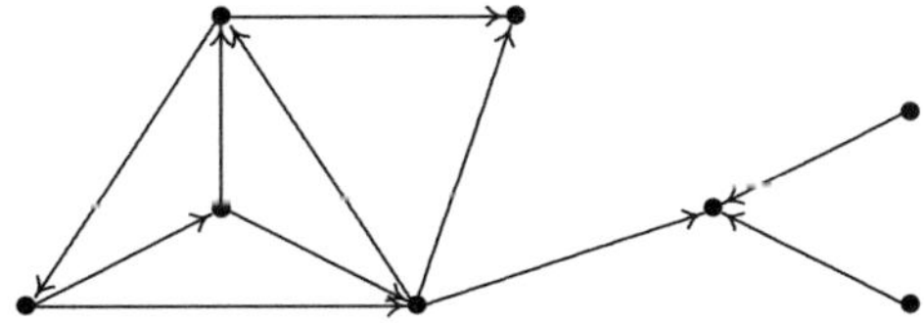

The graph Γ from (2.2.1) turned into a directed graph   (2.2.2)

Of course, every unweighted graph becomes a weighted one by assigning the weight 1 to every edge. An undirected graph with positive weights becomes a metric space by identifying each edge e with the interval of length $(w_e)^{-1}$. In particular, an unweighted graph then is a metric space where each edge is isometric to the unit interval. The distance between vertices then equals the length of the shortest path joining them. In particular, neighbors in the graph have distance 1.

We shall start with undirected and unweighted graphs as the simplest case. In the definition, we require that our graphs Γ be finite, a biologically directly plausible assumption. Moreover, we shall assume, unless stated to the contrary, that they are connected. That means that for every pair of distinct vertices i, j in Γ, there exists a path between them, that is, a sequence $i = i_0, i_1, \ldots, i_m = j$ of distinct vertices such that $i_{\nu-1} \sim i_\nu$ for $\nu = 1, \ldots, m$. Since we can decompose graphs that are not connected into their connected components, the connectivity assumption is no serious restriction.

An obvious way of representing a graph Γ with vertices $i = 1, \ldots, N$ is provided by its adjacency matrix $A = (a_{ij})$. In the unweighted case, we put $a_{ij} = 1$ when there is an edge from i to j and $= 0$ else. We have $a_{ii} = 0$ because we exclude self-loops of vertices, and Γ is undirected iff $a_{ij} = a_{ji}$ for all i, j. In the weighted case, we simply put $a_{ij} = w_{ij}$, the weight of the edge from i to j. Of course, most large graphs arising in applications are sparse, that is, between most pairs i, j, there is no edge. This means that most of the entries of the adjacency matrix are 0. Therefore, that matrix does not provide a very efficient way of encoding the graph. A more efficient way is provided by simply listing for each i those vertices that send links to i, together with the corresponding weights in the weighted case.

An isomorphism between graphs $\Gamma_1 = (V_1, E_1)$ and $\Gamma_2 = (V_2, E_2)$ is a bijection $\Phi : V_1 \to V_2$ that preserves neighborhood relations, that is, $i \sim j$ iff $\Phi(i) \sim \Phi(j)$. In other words, i and j are connected by an edge precisely if their images under Φ are. Isomorphisms preserve the degrees of vertices, that is, $n_i = n_{\Phi(i)}$ for every vertex i. An automorphism of Γ is an isomorphism from Γ onto itself. The identity map of the vertex set of Γ is obviously an automorphism, but there may or may not be others, depending on the structure of Γ. The automorphisms of Γ form a group under composition. We can then quantify the symmetry of Γ as the order of its automorphism group.

The number of graphs of order k grows very fast as a function of k, and therefore, it becomes unwieldy already for rather small k to list all graphs of order k. Therefore, it is of interest to develop constructions for particular classes or types of graphs. There exist deterministic and stochastic construction schemes. We shall discuss stochastic constructions below in 3.5 in the chapter on stochastic processes. Deterministic constructions typically produce rather regular graphs, that is ones with high degrees of symmetries whereas the stochastic constructions can produce typical representatives of larger classes of graphs. A paradigm of a symmetric graph is a complete graph, meaning that any two different vertices are connected by an edge. For a complete graph, every bijection of its vertices yields an automorphism, and therefore, it is maximally symmetric.

A cycle in Γ is a closed path $i_0, i_1, \ldots, i_m = i_0$ for which all the vertices $i_1, \ldots, i_m$ are distinct. For $m = 3$, we speak of a triangle. A cycle that contains all the vertices of Γ is called a Hamiltonian cycle (and such a cycle need not exist for a given graph). A graph without cycles is called a tree. A maximal tree contained in a graph Γ is called a spanning tree. A spanning tree is obtained by eliminating all cycles from a graph, that is, by cutting an edge in each cycle.

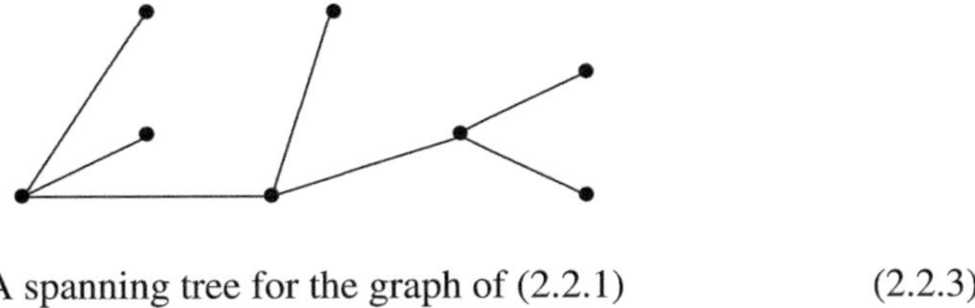

A spanning tree for the graph of (2.2.1)   (2.2.3)

A graph is called k-regular if all vertices have the same degree k. As already mentioned, a graph is bipartite if its vertex set can be decomposed into two disjoint components V_1, V_2 such that whenever $i \sim j$, then i and j are in different components.

A bipartite graph   (2.2.4)

It is not hard to see that a graph is bipartite iff it does not contain cycles of odd length. In particular, it cannot contain any triangles.

Another useful concept for analyzing graphs is the k-core. For $k \in \mathbb{N}$, the k-core of a graph Γ is the not necessarily connected maximal subgraph H of Γ with the property that every vertex of H has at least k neighbors in H, that is, its degree in H is at least k. When we exclude the trivial case of an isolated vertex, then Γ itself coincides with its 1-core. When Γ is a tree, already its 2-core is empty. Every cycle of Γ is contained in its 2-core. The core decomposition of Γ, that is, the successive determination of its k-cores for increasing k, is a computationally simple way of decomposing the graph.

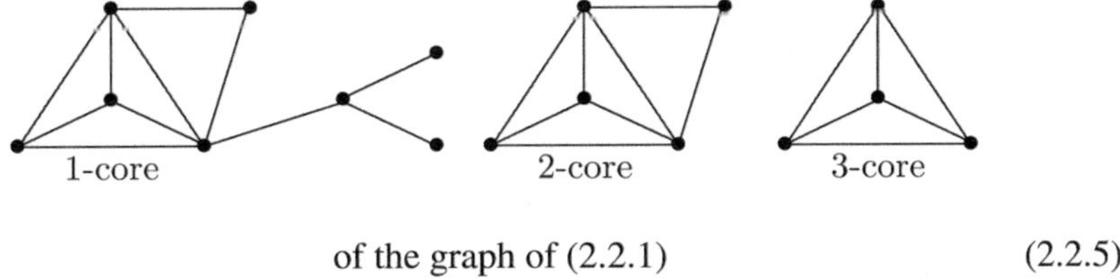

of the graph of (2.2.1)   (2.2.5)

There exist other parameters that describe certain—more or less—important qualitative properties of graphs. One set of such parameters arises from the metric on the graph generated by the above assignment of length 1 to every edge. The diameter of the graph is the maximal distance between any two of its nodes. As an example how such a parameter can distinguish between typical and non-typical, special graphs, we record that there exists a constant c with the property that the fraction of all graphs with N nodes having diameter exceeding $c \log N$ tends to 0 for $N \to \infty$. Informally expressed, most graphs of N nodes have a diameter of order $\log N$. Thus, graphs with large diameters, like a chain $i_1 \sim i_2 \sim \cdots \sim i_N$ with no other edges, are rare. In the other direction, that is, considering graphs with very small diameters, of course, a fully connected graph has diameter 1. However, one can realize a small diameter already with much fewer edges; namely, one selects one central node to which every other node is connected. In that manner, one obtains a graph of N nodes with $N-1$ edges and diameter 2. (This graph is called the $(N-1)$-star, and it will be discussed further below.) Of course, the central node then has a very large degree, namely $N-1$. It is a big hub. Similarly, one can construct graphs with a few hubs, so that none of them has to be quite that big, efficiently distributed so that the diameter is still rather small. Such graphs can be realized as so-called scale free graphs to be discussed below. Another useful quantity is the average distance between nodes in the graph. The property of having a small diameter or average distance has been called the small-world effect.

A rather different quantity is the clustering coefficient that measures how many connections there exist between the neighbors of nodes. For this purpose, a triple is a set of three connected vertices, that is, a path with three vertices. As mentioned, a cycle of length 3 is called a triangle

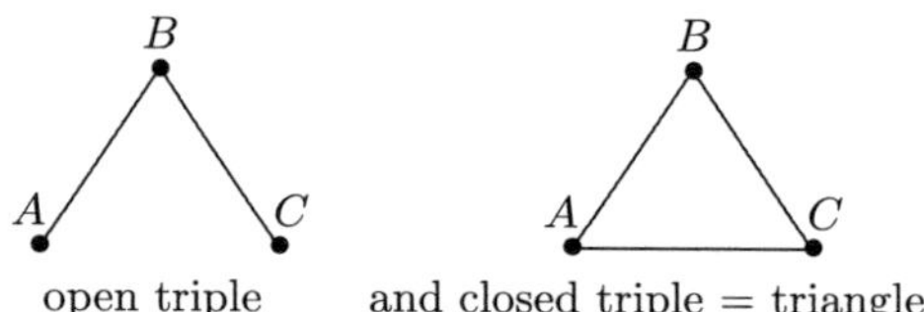

Note that the triangle contains three triples, ABC, BCA and CAB.
The (global) clustering coefficient then is defined as

$$C := \frac{3 \times \text{number of triangles}}{\text{number of connected triples of nodes}}. \qquad (2.2.6)$$

The normalization is that C becomes one for a fully connected graph. It vanishes for trees and other bipartite graphs.
As already mentioned, a k-star is a graph consisting of one central vertex connected to k peripheral vertices, with no connections between those other vertices.

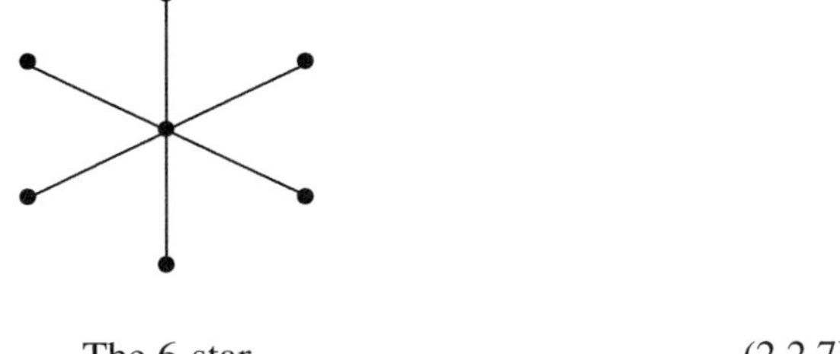

The 6-star

(2.2.7)

In particular, each k-star is a tree. A k-star has $\binom{k}{2}$ connected triples of nodes, obtained by connecting the central node with any two peripheral ones. Thus, when we want to compute the clustering index of a graph, we count $\sum_{i \in V} \binom{n_i}{2}$ connected triples of vertices. Thus, the graph Γ of (2.2.1) has 5 triangles and 26 connected triples of nodes, and hence its clustering coefficient is $\frac{15}{26}$.

A triangle is a cycle of length 3. One may then also count the number of cycles of length k, for integers >3. A different generalization consists in considering complete subgraphs of order k. Here, the complete k-graph K_k is the graph with k vertices and links between all $i \neq j$. A k-clique in a graph Γ is a subgraph that is a complete k-graph. For example, for $k = 4$, we would have a subset of 4 nodes that are all mutually connected.

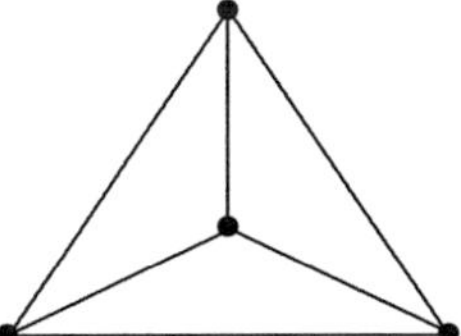

The complete graph K_4, the only 4-clique in the graph of (2.2.1)

(2.2.8)

One may then associate a simplicial complex to our graph by assigning a k-simplex to every such complete subgraph, with obvious incidence relations. For example, two such k-simplices share a $(k-1)$-dimensional face and are called adjacent when the two corresponding complete k-subgraphs have a complete $(k-1)$-graph in common. This is the basis of topological combinatorics, enabling one to apply tools from simplicial topology to graph theory. See for instance [65].

Besides the complete graphs K_k, one also frequently encounters the complete bipartite graphs $K_{m,n}$ consisting of two classes of m and n, resp., vertices such that every vertex in the first class is connected with every vertex in the second class.

The complete bipartite graph $K_{2,3}$

(2.2.9)

The graph $K_{1,n}$ is of course simply the n-star.

A basic question in the analysis of graphs is the cluster decomposition. That means that we try to find subgraphs, the clusters, that are densely connected inside, but only sparsely connected to the rest of the graph. For example, one can try to disconnect the graph by cutting as few edges as possible, to obtain two large (super) clusters,

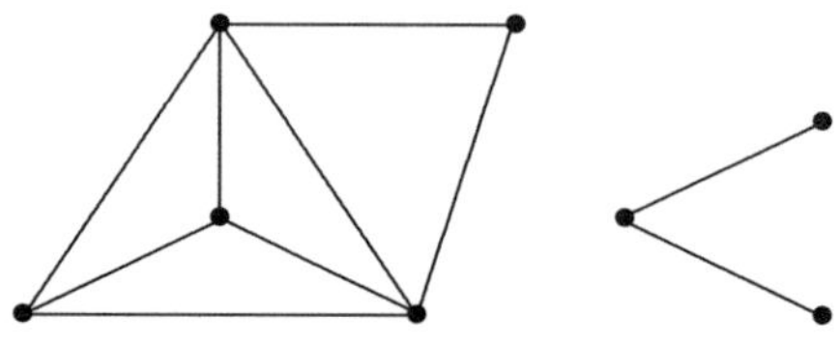

$$\text{A decomposition of the graph } \Gamma \text{ from (2.2.1)} \qquad (2.2.10)$$

and then perhaps iterate the process inside these superclusters to find a finer decomposition. Conversely, one can try to build up the clusters from inside, for example by identifying maximal sets of adjacent k-cliques, or, equivalently, in the simplicial complex defined above, finding maximal sets of k-simplices that are connected by $(k-1)$-dimensional faces. Here, the clusters found are typically not disjoint, in contrast to those obtained by the edge-cutting methods. Of course, one may then analyze the overlap between those clusters.

Concerning the number of edges needed to disconnect a graph, some insight is provided by the following result of Menger:

Lemma 2.2.1. *Let V_1 and V_2 be disjoint subsets of the vertex set of a graph $\Gamma = (V, E)$. The minimal number of edges that need to be deleted from Γ in order to disconnect it in such a manner that V_1 and V_2 are in different components is equal to the maximum number of edge-disjoint paths (that is no two paths are allowed to have an edge in common, even though they may well pass through the same vertex) with one endpoint in V_1 and the other in V_2.*

Another general question is to identify the most important "core" of the graph. The k-core defined above is one useful concept for that. The idea there is that a node is important when it is connected with other important nodes. Thus, one finds the core by successively deleting the less important nodes. That procedure might make some nodes that have originally been highly connected, that is, have a large degree, less relevant, because they had only been connected to other nodes of low degrees. Therefore, in particular, the degree of a node in general is not a good measure of its importance. One can also quantify the importance of a vertex or an edge by counting how many shortest connections between pairs of nodes pass through them. Again, one should be a bit cautious here because in some cases, there exist alternatives to shortest paths that are not substantially longer but that avoid the vertex or edge in question. In other words, sometimes vertices or edges can easily be replaced as parts of short connections while in other cases that may not possible. When one decides the importance according to such considerations, this effect should also be taken into account.

2.2.3 The Graph Laplacian and its Spectrum

As before, Γ is a finite and connected graph. Probably the most powerful and comprehensive set of invariants comes from the spectrum of the graph Laplacian of Γ to which we now turn. (In general terms, this means that, in order to analyze a graph Γ, we shall study functions defined on Γ. These functions will then be decomposed in terms of a particular set of basis functions, as in Fourier analysis. From those basis functions, we shall obtain spectral values that incorporate the characteristic properties of Γ.)

There are several non-equivalent definitions of the graph Laplacian employed in the literature. In order to clarify this issue, we assign weights $b_i\,(>0)$ to the vertices[1] and introduce an L^2-product for (complex-valued) functions on Γ:

$$(u, v) := \sum_{i \in V} b_i u(i)\overline{v(i)}. \tag{2.2.11}$$

(Since we shall only consider real operators below, it suffices to consider real valued functions, and then the complex conjugate in (2.2.11) is not relevant.)

The most natural choices are $b_i = 1$ or $b_i = n_i$ where n_i is the degree of the vertex i.[2] We may then choose an orthonormal base of that space $L^2(\Gamma)$. In order to find such a basis that is also well adapted to dynamical aspects, we study the graph Laplacian

$$\Delta : L^2(\Gamma) \to L^2(\Gamma)$$

$$\Delta v(i) := \frac{1}{b_i}\left(\sum_{j,j\sim i} v(j) - n_i v(i) \right) \tag{2.2.12}$$

where $j \sim i$ means that j is a neighbor of i.[3]

---

[1] These vertex weights should not be confused with the edge weights discussed above; in other words, here, we are *not* considering weighted graphs in the sense defined above.

[2] For purposes of normalization, one might wish to put an additional factor N in front of the product where N is the number of vertices of the graph or, equivalently, divide all the vertex weights by N, but we have decided to omit that factor in our conventions.

[3] There are several different definitions of the graph Laplacian in the literature. Some of them are equivalent to ours inasmuch as they yield the same spectrum, but others are not. The reason is simply that the weights in the underlying product are chosen differently. The operator $Lv(i) := n_i v(i) - \sum_{j,j\sim i} v(j)$ that is often employed in the literature corresponds to the weights $b_i = 1$ (up to the minus sign, of course). The operator $\mathcal{L}v(i) := v(i) - \sum_{j,j\sim i} \frac{1}{\sqrt{n_i}\sqrt{n_j}} v(j)$ employed in the

We, in contrast to much of the literature on graph theory (see e.g. [50]), but in accordance with [28], prefer the weights $b_i = n_i$ over $b_i = 1$ because the former are well adapted to random walks and conservation laws. (When we have a particle randomly moving on a graph with step size 1 then when it is at vertex i it can choose each of the neighbors of i with probability $1/n_i$ for its next move, and this leads to the corresponding factor in the Laplace operator underlying that random walk. This process will be investigated in detail in Sect. 4.2.1.)

The idea behind the definition of Δ is of course that one compares the value of a function v at a vertex i with the average of the values at the neighbors of i. When that average is larger than the value at i, we have $(\Delta v)(i) > 0$.

The important properties of Δ are the following ones:

1. Δ is selfadjoint w.r.t. $(.,.)$:

$$(u, \Delta v) = (\Delta u, v) \tag{2.2.13}$$

for all $u, v \in L^2(\Gamma)$.[4] This holds because the neighborhood relation is symmetric.
2. Δ is nonpositive:

$$(\Delta u, u) \leq 0 \tag{2.2.14}$$

for all u. This follows from the Cauchy-Schwarz inequality.
3. $\Delta u = 0$ precisely when u is constant. This one sees by observing that, when $\Delta u = 0$, there can neither be a vertex i with $u(i) \geq u(j)$ for all $j \sim i$ with strict inequality for at least one such j, that is, a nontrivial local maximum, nor a nontrivial local minimum, as this would contradict the fact that $\Delta u(i) = 0$ means that the value $u(i)$ is the average of the values at the neighbors of i. Since Γ is connected, u then has to be a constant (when Γ is not connected, a solution of $\Delta u = 0$ is constant on every connected component of Γ.)

The preceding properties have consequences for the eigenvalues of Δ:

- By 1, the eigenvalues are real.
- By 2, they are nonpositive. We write them as $-\lambda_k$ so that the eigenvalue equation becomes[5]

$$\Delta u_k + \lambda_k u_k = 0. \tag{2.2.15}$$

(Footnote 3 continued)

monograph [28], apart from the minus sign, has the same eigenvalues as Δ for the weights $b_i = n_i$: if $\Delta v(i) = \mu v(i)$, then $w(i) = \sqrt{n_i} v(i)$ satisfies $\mathcal{L}w(i) = -\mu w(i)$.

[4] An operator $A = (A_{ij})$ is symmetric w.r.t. a product $\langle v, w \rangle := \sum_i b_i v(i)\overline{w(i)}$, that is, $\langle Av, w \rangle = \langle v, Aw \rangle$ if $b_i A_{ij} = b_j A_{ji}$ for all indices i, j. The b_i are often called multipliers in the literature.

[5] Subsequently, we shall thus call the λ_k instead of the $-\lambda_k$ the eigenvalues, in order to avoid negative quantities. The sign problem comes from the—traditional—definition of the Laplacian (2.2.12) as a nonpositive operator.

- By 3, the smallest eigenvalue is $\lambda_0 = 0$. Since we assume that Γ is connected, this eigenvalue is simple, that is

$$\lambda_k > 0 \qquad\qquad (2.2.16)$$

for $k > 0$ where we order the eigenvalues as

$$\lambda_0 = 0 < \lambda_1 \leq \dots \leq \lambda_K$$

where we put $K := N - 1$.

We next consider, for neighbors i, j,

$$Du(i, j) := u(i) - u(j). \qquad\qquad (2.2.17)$$

D can be considered as a map from functions on the vertices of Γ to functions on the edges of Γ. In order to make the latter space also an L^2-space, we introduce the product

$$(Du, Dv) := \sum_{e=(i,j)} (u(i) - u(j))(v(i) - v(j)). \qquad\qquad (2.2.18)$$

Note that we are summing here over edges, and not over vertices. If we did the latter, we would need to put in a factor $1/2$ because each edge would then be counted twice. We also point out that in contrast to the product of (2.2.11), $(u, v) = \sum_i b_i u(i) v(i)$, we do not include weights here. The reason is that here the sum should be considered as a sum of edges and not one over vertices, and since we are considering unweighted graphs at this point, the edges do not carry any natural weights.

The product (2.2.18) encodes more information about the graph than the product (2.2.11). The latter only depends on the weights, but not on the connection structure of the graph. There exist many structurally quite diverse graphs with the same weight sequence, and given a graph, one can rewire it by a cross exchange of edges without changing the degrees of the nodes. Namely, given vertices $i_1 \sim j_1$ and $i_2 \sim j_2$, but without edges between i_1 and i_2, nor between j_1 and j_2, we create a new graph by deleting the edges between i_1 and j_1 and between i_2 and j_2 and inserting new edges between i_1 and i_2 and between j_1 and j_2. That operation preserves the degrees of all vertices, and therefore also the product (2.2.11) for any functions u, v on the graph. (2.2.18), in contrast, is affected because the edge set is changed.

We have

$$(Du, Dv) = \frac{1}{2} \sum_i (n_i u(i)v(i) + \sum_j n_j u(j)v(j) - 2 \sum_{j \sim i} u(i)v(j))$$

$$= - \sum_i u(i) \sum_{j \sim i} (v(j) - v(i))$$

$$= -(u, \Delta v). \tag{2.2.19}$$

Thus, our product (2.2.18) is naturally related to the Laplacian Δ.

We may find an orthonormal basis of $L^2(\Gamma)$ consisting of eigenfunctions of Δ,

$$u_k, \ k = 0, ..., K$$

($K = N - 1$). This is achieved as follows. We iteratively define, with $H_0 := H := L^2(\Gamma)$ being the Hilbert space of all real-valued functions on Γ with the scalar product $(., .)$,

$$H_k := \{v \in H : (v, u_i) = 0 \text{ for } i \leq k - 1\}, \tag{2.2.20}$$

starting with a constant function u_0 as the eigenfunction for the eigenvalue $\lambda_0 = 0$. Also

$$\lambda_k := \inf_{u \in H_k - \{0\}} \frac{(Du, Du)}{(u, u)}, \tag{2.2.21}$$

that is, we claim that the eigenvalues can be obtained as those infima. First of all, since $H_k \subset H_{k-1}$, we have

$$\lambda_k \geq \lambda_{k-1}. \tag{2.2.22}$$

Secondly, since the expression in (2.2.21) remains unchanged when a function u is multiplied by a nonzero constant, it suffices to consider those functions that satisfy the normalization

$$(u, u) = 1 \tag{2.2.23}$$

whenever convenient.

We may find a function u_k that realizes the infimum in (2.2.21), that is

$$\lambda_k = \frac{(Du_k, Du_k)}{(u_k, u_k)}. \tag{2.2.24}$$

Since then for every $\varphi \in H_k, t \in \mathbb{R}$

$$\frac{(D(u_k + t\varphi), D(u_k + t\varphi))}{(u_k + t\varphi, u_k + t\varphi)} \geq \lambda_k, \tag{2.2.25}$$

the derivative of that expression w.r.t. t vanishes at $t = 0$, and we obtain, using (2.2.19)

$$0 = (Du_k, D\varphi) - \lambda_k(u_k, \varphi) = -(\Delta u_k, \varphi) - \lambda_k(u_k, \varphi) \tag{2.2.26}$$

for all $\varphi \in H_k$; in fact, this even holds for all $\varphi \in H$, and not only for those in the subspace H_k, since for $i \leq k - 1$

$$(u_k, u_i) = 0 \tag{2.2.27}$$

and

$$(Du_k, Du_i) = (Du_i, Du_k) = -(\Delta u_i, u_k) = \lambda_i(u_i, u_k) = 0 \tag{2.2.28}$$

since $u_k \in H_k$. Thus, if we also recall (2.2.19),

$$(\Delta u_k, \varphi) + \lambda_k(u_k, \varphi) = 0 \tag{2.2.29}$$

for all $\varphi \in H$ whence

$$\Delta u_k + \lambda_k u_k = 0. \tag{2.2.30}$$

Since, as noted in (2.2.23), we may require

$$(u_k, u_k) = 1 \tag{2.2.31}$$

for $k = 0, 1, \ldots, K$ and since the u_k are mutually orthogonal by construction, we have constructed an orthonormal basis of H consisting of eigenfunctions of Δ. Thus we may expand any function f on Γ as

$$f(i) = \sum_k (f, u_k) u_k(i). \tag{2.2.32}$$

We then also have

$$(f, f) = \sum_k (f, u_k)^2 \tag{2.2.33}$$

since the u_k satisfy

$$(u_j, u_k) = \delta_{jk}, \tag{2.2.34}$$

the condition for being an orthonormal basis. Finally, using (2.2.33) and (2.2.19), we obtain

$$(Df, Df) = \sum_k \lambda_k (f, u_k)^2. \tag{2.2.35}$$

We next state **Courant's minimax principle**:
Let P^k be the collection of all k-dimensional linear subspaces of H. We have

$$\lambda_k = \max_{L \in P^k} \min \left\{ \frac{(Du, Du)}{(u, u)} : u \neq 0, \ (u, v) = 0 \text{ for all } v \in L \right\} \tag{2.2.36}$$

and dually

$$\lambda_k = \min_{L \in P^{k+1}} \max \left\{ \frac{(Du, Du)}{(u, u)} : u \in L \setminus \{0\} \right\}. \tag{2.2.37}$$

In words: In (2.2.36), we consider the minimal Rayleigh quotient under k constraints, and we maximize that w.r.t. the constraints. In (2.2.37), we consider the maximal Rayleigh quotient for $k + 1$ degrees of freedom, and we minimize that w.r.t. those degrees of freedom.
To verify these relations, we recall (2.2.21)

$$\lambda_k = \min \left\{ \frac{(Du, Du)}{(u, u)} : u \neq 0, (u, u_j) = 0 \text{ for } j = 0, ..., k - 1 \right\}. \tag{2.2.38}$$

Dually, we have

$$\lambda_k = \max \left\{ \frac{(Du, Du)}{(u, u)} : u \neq 0 \text{ linear combination of } u_j \text{ with } j \leq k \right\}. \tag{2.2.39}$$

The latter maximum is realized when u is a multiple of the kth eigenfunction, and so is the minimum in (2.2.38). If now L is any $k + 1$-dimensional subspace, we may find some v in L that satisfies the k conditions

$$(v, u_j) = 0 \text{ for } j = 0, ..., k - 1. \tag{2.2.40}$$

From (2.2.33) and (2.2.35), we then obtain

$$\frac{(Dv, Dv)}{(v, v)} = \frac{\sum_{j \geq k} \lambda_j (v, u_j)^2}{\sum_{j \geq k} (v, u_j)^2} \geq \lambda_k. \tag{2.2.41}$$

This implies

$$\max_{v \in L \setminus \{0\}} \frac{(Dv, Dv)}{(v, v)} \geq \lambda_k. \tag{2.2.42}$$

We then obtain (2.2.37). Equation (2.2.36) follows in a dual manner. In particular, for any eigenfunction u for some eigenvalue $\lambda \neq 0$, we then have

$$\lambda = \frac{(Du, Du)}{(u, u)} \tag{2.2.43}$$

For a fully connected graph,[6] when all the weights b_i are equal, also all the nontrivial eigenvalues are equal. For our preferred choice of weights, $b_i = n_i (= N - 1$ for a fully connected graph of N vertices), we have

$$\lambda_1 = \ldots = \lambda_K = \frac{N}{N - 1} \tag{2.2.44}$$

since

$$\Delta v = -\frac{N}{N - 1} v \tag{2.2.45}$$

for any v that is orthogonal to the constants, that is

$$\frac{1}{N} \sum_{i \in V} n_i v(i) = 0. \tag{2.2.46}$$

In more detail, for a fully connected graph of N vertices, for v satisfying (2.2.46),

$$\begin{aligned}
\Delta v(i) &= \frac{1}{n_i} \sum_{j, j \sim i} v(j) - v(i) \\
&= \frac{1}{N - 1} \sum_{j \neq i} v(j) - v(i) \\
&= \left(-\frac{1}{N - 1} - 1\right) v(i) \text{ since by (2.2.46) } v(i) = -\sum_{j \neq i} v(j) \\
&= -\frac{N}{N - 1} v(i).
\end{aligned}$$

We also recall that since Γ is connected, the trivial eigenvalue $\lambda_0 = 0$ is simple. If Γ had two components, then the next eigenvalue λ_1 would also become 0. A corresponding eigenfunction would be equal to a constant on each component, the two values chosen such (2.2.46) is satisfied; in particular, one of the two would be positive, the other one negative. We therefore expect that for graphs with a pronounced community structure, that is, for ones that can be broken up into two large components by deleting only few edges as discussed above, the eigenvalue λ_1 should be close to 0. Formally, this is easily seen from the variational characterization

---

[6] A fully connected graph is a complete graph K_N, possibly with vertex weights.

$$\lambda_1 = \min\{\frac{\sum_{e=(i,j)\in E}(v(i) - v(j))^2}{\sum_i b_i v(i)^2} : \sum_i b_i v(i) = 0\} \tag{2.2.47}$$

(see (2.2.21) and observe that $\sum_i b_i v(i) = 0$ is equivalent to $(v, u_0) = 0$ as the eigenfunction u_0 is constant). Namely, if two large components of Γ are only connected by few edges, then one can make v constant on either side, with opposite signs so as to respect the normalization (2.2.46) with only a small contribution from the numerator.

More generally, when Γ consists of several clusters with only very few connections between them, one should find several eigenvalues close to 0.

The strategy for obtaining an eigenfunction for the first eigenvalue λ_1 is, according to (2.2.47), to do the same as one's neighbors. Because of the constraint $\sum_i b_i v(i) = 0$, this is not globally possible, however. The first eigenfunction thus exhibits oscillations with the lowest possible frequency . Thus, if we take such a first eigenfunction u_1 and consider the connected components that remain after deleting all edges at whose endpoints u_1 has different signs, then there are precisely two such components, one on which u_1 is positive and one on which it is negative. More generally, the number of connected components of Γ where an eigenfunction for the kth eigenvalue has a fixed sign is at most $k + 1$ when the eigenvalues are ordered in increasing order and appropriately when they are not simple, according to a version of Courant's nodal domain theorem proved by Gladwell-Davies-Leydold-Stadler [48].

We once more consider the case $b_i = n_i$. As noted, for a complete graph, we have $\lambda_1 = \frac{N}{N-1}$, see (2.2.44). For any other graph, that is, for any graph that is not complete, we have

$$\lambda_1 \leq 1. \tag{2.2.48}$$

This follows from (2.2.47), by taking two vertices i_1, i_2 that are not connected by an edge and by assigning values of u to those points satisfying $n_{i_1} u(i_1) + n_{i_2} u(i_2) = 0$ and 0 to all other vertices. The quotient in (2.2.47) then becomes 1, and therefore, the infimum characterizing λ_1 has to be ≤ 1.

By way of contrast, according to (2.2.37), the highest eigenvalue is given by

$$\lambda_K = \max_{u \neq 0} \frac{(Du, Du)}{(u, u)}. \tag{2.2.49}$$

Thus, the strategy for obtaining an eigenfunction for the highest eigenvalue is to do the opposite what one's neighbors are doing, for example to assume the value 1 when the neighbors have the value -1. Thus, the corresponding eigenfunction will exhibit oscillations with the highest possible frequency. Here, the obstacle can be local. Namely, any triangle, that is, a triple of three mutually connected nodes, presents such an obstacle. More generally, any cycle of odd length makes an alternation of the values 1 and -1 impossible. The optimal situation here is represented by a bipartite graph, that is, a graph that consists of two sets Γ_+, Γ_- of nodes without any links between nodes in the same such subset. Thus, one can put $u_K = \pm 1$ on $\Gamma_\pm$. For our choice $b_i = n_i$, which we shall now adopt for the subsequent discussion, one then finds

$$\lambda_K = 2 \tag{2.2.50}$$

for a bipartite graph.

In contrast, the highest eigenvalue λ_K becomes smallest on a fully connected graph, namely

$$\lambda_K = \frac{N}{N-1} \tag{2.2.51}$$

according to (2.2.46). For graphs that are neither bipartite nor fully connected, this eigenvalue lies strictly between those two extremal possibilities.

Perhaps the following caricature can summarize the preceding: For minimizing λ_1— the minimal value being 0—one needs two subsets that can internally be arbitrarily connected, but that do not admit any connection between each other. For maximizing λ_K—the maximal value being 2—one needs two subsets without any internal connections, but allowing arbitrary connections between them. In either situation, the worst case—that is, a maximal value for λ_1 and a minimal value for λ_K—is represented by a fully connected graph. In fact, in that case, λ_1 and λ_K coincide.

Let us consider bipartite graphs in some more detail. We already noted above that on a bipartite graph, we can determine the highest eigenfunction u_K explicitly, as ± 1, being $+1$ on one set, -1 on the other set of vertices defining the bipartition. In fact, it is clear from that construction that this property is equivalent to the bipartiteness of the graph. Actually, if the graph is bipartite, then even more is true: Whenever λ_k is an eigenvalue, then so is $2 - \lambda_k$. Since 0 is an eigenvalue for any graph, this criterion implies our observation that 2 is an eigenvalue. The general statement is not difficult to see: Let G_1, G_2 be the two vertex sets defining the bipartition. When u_k is an eigenfunction for the eigenvalue λ_k, then

$$\tilde{u}_k(i) := \begin{cases} u_k(i) \text{ for } i \in G_1 \\ -u_k(i) \text{ for } i \in G_2 \end{cases} \tag{2.2.52}$$

is an eigenfunction with eigenvalue $2 - \lambda_k$ as is readily verified.

We now present some results from [15] about controlling the highest eigenvalue. In order to understand the significance of the highest eigenvalue λ_K better, we now derive some general identity first, for a function u on the vertex set of Γ.

$$\sum_i \frac{1}{n_i} \sum_{j,k,\, j\sim i, k\sim i} (u(j) - u(k))^2$$

$$= \sum_i \sum_{k,\, k\sim i} \frac{1}{n_i} \sum_{j,\, j\sim i} (u(j) - u(k))^2$$

$$= \sum_i \left(\sum_{k,k \sim i} \left(\frac{1}{n_i} \sum_{j,j \sim i} u(j)^2 - \frac{2}{n_i} \sum_{j,j \sim i} u(j)u(k) + u(k)^2 \right) \right)$$

$$= \sum_i \left(2 \sum_{j,j \sim i} u(j)^2 - \frac{2}{n_i} (\sum_{j,j \sim i} u(j))^2 \right)$$

$$= 2 \sum_i \sum_{j,j \sim i} u(j)^2 - \sum_i 2n_i \left(\frac{1}{n_i} \sum_{j,j \sim i} u(j) \right)^2.$$

We now observe that we can replace u by $u - u(i)$ in the first and hence also in all subsequent lines. This yields

$$\sum_i \frac{1}{n_i} \sum_{j,k,j \sim i,k \sim i} (u(j) - u(k))^2$$

$$= 2 \sum_i \sum_{j,j \sim i} (u(j) - u(i))^2 \quad \sum_i 2n_i \left(\frac{1}{n_i} \sum_{j,j \sim i} (u(j) - u(i)) \right)^2$$

$$= 2 \sum_i \sum_{j,j \sim i} (u(j) - u(i))^2 - \sum_i 2n_i (\Delta u(i))^2.$$

When u now is an eigenfunction, $\Delta u + \lambda u = 0$ for some eigenvalue λ, then, recalling (2.2.43), we obtain

$$\sum_i \frac{1}{n_i} \sum_{j,k,j \sim i,k \sim i} (u(j) - u(k))^2 = 2\lambda(2 - \lambda) \sum_i n_i u(i)^2. \tag{2.2.53}$$

Using (2.2.43) again, we can also reformulate this as

$$2 - \lambda = \frac{\sum_i \frac{1}{n_i} \sum_{j,k,j \sim i,k \sim i} (u(j) - u(k))^2}{\sum_i \sum_{j,j \sim i} (u(j) - u(i))^2}. \tag{2.2.54}$$

We now want to employ (2.2.54) to interpret $2 - \lambda_K$ (λ_K being the largest eigenvalue of our graph) as quantifying how much Γ is locally different from being bipartite, recalling that this quantity is 0 precisely if Γ happens to be bipartite.

In order to develop some intuition, we start with a bipartite graph Γ_0 with M vertices. We consider a highest eigenfunction $\bar{u}$ that is $+1$ on one class and -1 on the other class of vertices, as described above. In particular,

$$\frac{\frac{1}{2} \sum_{j \sim k} (\bar{u}(j) - \bar{u}(k))^2}{\sum_i n_i \bar{u}(i)^2} = 2. \tag{2.2.55}$$

We add another vertex i_0 and connect it to one of the edges of Γ_0. Of course, this new graph Γ_1 then is again bipartite, but we extend $\bar{u}$ by $\bar{u}(i_0) = 0$ to Γ_1. Thus, the numerator and the denominator of (2.2.55) are both increased by 1. Given any small $\epsilon > 0$, by assuming that Γ_0 is sufficiently large, that is, $\sum_i n_i$ is sufficiently large, we can therefore achieve that, for Γ_1,

$$\frac{\frac{1}{2}\sum_{j\sim k}(\bar{u}(j) - \bar{u}(k))^2}{\sum_i n_i \bar{u}(i)^2} > 2 - \epsilon. \tag{2.2.56}$$

Now, this is not affected when we construct a graph Γ by attaching another graph Γ_2 at i_0 and extend $\bar{u}$ by 0 to all of Γ_2. For instance, Γ_2 could be a complete graph of N vertices, for any N. In particular, the difference $2 - \lambda_K$ (where λ_K is the largest eigenvalue of Γ) which has to be larger than $2 - \epsilon$ by (2.2.51), is not very sensitive to the shape of Γ_2. This implies, for instance, that $2 - \lambda_K$ cannot reflect a global quantity like the clustering coefficient C of (2.2.6) that expresses an averaged difference from a graph being bipartite. In fact, our construction of attaching a complete graph K_N to a bipartite graph Γ_0 through a connecting node produces a graph with C arbitrarily close to its maximal value 1 when N is sufficiently large. By extending this example, we can also see that we should have many eigenvalues λ for which $2 - \lambda$ is small when the graph possesses several relatively large bipartite or almost bipartite parts that are only loosely connected with the rest. This is analogous to the fact that a graph possesses several small eigenvalues when it has many relatively large components that are only loosely connected to the rest, that is, when the graph can be easily decomposed into several large clusters. Of course, for a nonconnected graph, that is, one with several components without links between them, the spectrum simply is the union of the spectra of the components. Therefore, by the continuity principle, a graph consisting of clusters that are only loosely connected to each other has its spectrum approximated (in a sense not made completely precise here) by the spectra of these clusters, that is, by that of the graph resulting from deleting the few links between the clusters.

We can use (2.2.54), however, in order to control $2 - \lambda_K$ by the following local clustering measure

$$C_0(\Gamma) := \max\{\alpha : \text{ for each } i \in \Gamma, \text{ at least } \alpha n_i \text{ of its edges are contained in some triangle}\}. \tag{2.2.57}$$

Again, $C_0 = 0$ for a bipartite and $C_0 = 1$ for a complete graph. Thus, let us analyze (2.2.54) with this quantity in mind. We want to control $2 - \lambda_K$ from below in terms of C_0. This means that we need to match any term $(u(i) - u(j))^2$ in the denominator by some term in the numerator of comparable magnitude. Now, given such a term, we have two possibilities. Either we can at least find $\frac{\alpha n_i}{2}$ neighbors k of i for which $(u(k) - u(j))^2 \geq \frac{1}{2}(u(i) - u(j))^2$. Then $(u(i) - u(j))^2$ is matched in the numerator. Or, for at least $\frac{\alpha n_i}{2}$ neighbors k of i for which the edge $e = (i, k)$ is contained in some triangle (i, k, ℓ), we have $(u(i) - u(k))^2 \geq \frac{1}{12}(u(i) - u(j))^2$ for at least αn_i neighbors

k that are contained in some triangle (i, k, ℓ). Therefore, taking one of those k and one such triangle (i, k, ℓ), for every other vertex m of ℓ, either $(u(i) - u(m))^2$ or $(u(k) - u(m))^2$ has to be sufficiently large. Since we have the choice between at least αn_i such vertices k, all the edges $e = (i, j)$ can thus be matched with a controlled amount of duplication. Thus, $2 - \lambda_K$ can be controlled from below in terms of C_0, or conversely, C_0 can be controlled from above in terms of $2 - \lambda_K$. The control in the other direction does not work quite, because $2 - \lambda_K$ can still be made relatively large in a graph like that obtained from the complete graph K_N by attaching another node i_0 with a single connection to one of the vertices of K_N. Here, $C_0 = 0$, because the only edge from i_0 is not contained in a triangle. Perhaps more a more important example is a graph with many cycles of odd length, but all of them of length at least 5. Here, $C_0 = 0$ as there are no triangles, but $2 - \lambda_K \neq 0$ because the graph is not bipartite as bipartite graphs can only have cycles of even length.

In passing, we also observe that by a reasoning similar to that for (2.2.53), we can also show

$$\sum_i \sum_{k, k \sim i} \left(\frac{1}{n_i} \sum_{j, j \sim i} (u(j) - u(k)) \right)^2 = \lambda(2 - \lambda) \sum_i n_i u(i)^2. \qquad (2.2.58)$$

Again, this can be used to estimate the local difference from being bipartite in terms of $2 - \lambda_K$.

In fact, the preceding constructions can be understood and significantly extended through the concept of a neighborhood graph, see [15].

Having looked at the smallest and largest eigenvalues, we now take a look at the one in the middle, $\lambda = 1$ (we again fix the weights in (2.2.12) to be n_i). In order to see that this eigenvalue is special, we rewrite the eigenvalue equation for $\lambda = 1$ as

$$0 = \Delta v(i) + v(i) = \frac{1}{n_i} \left(\sum_{j, j \sim i} v(j) - n_i v(i) \right) + v(i) = \frac{1}{n_i} \sum_{j, j \sim i} v(j). \qquad (2.2.59)$$

Thus, an eigenfunction for the eigenvalue 1 is *balanced* in the sense that for every node, the average of the values of its neighbors vanishes.

There is a simple way to generate the eigenvalue 1: node duplication. That means that we take some graph Γ_0 and some node $i_0 \in \Gamma_0$ and create a new graph Γ by adjoining an additional node j_0 to Γ_0 by the prescription that j_0 gets connected to the same nodes as i_0. That is, whenever a node $i \in \Gamma_0$ is connected to i_0, then we also connect it to j_0. Note that i_0 and j_0 are not directly connected by this rule. The node j_0 can then be considered as the double of i_0 because it shares the same neighbors in Γ.

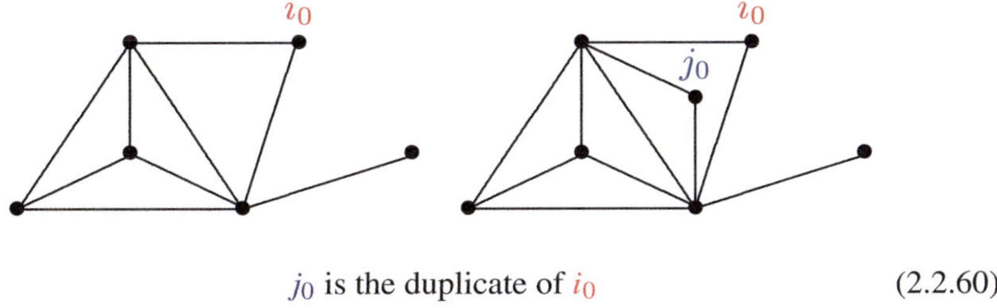

j_0 is the duplicate of i_0 (2.2.60)

We now simply observe that the function

$$u_1(i) := \begin{cases} 1 \text{ for } i = i_0 \\ -1 \text{ for } i = j_0 \\ 0 \text{ else} \end{cases} \qquad (2.2.61)$$

satisfies the eigenvalue equation (2.2.59). By repeated node duplication, we can thus generate a graph with an arbitrary high multiplicity of the eigenvalue 1. In particular, a complete bipartite graph $K_{m,n}$, that is, a bipartite graph consisting of one class of size m and another of size n, such that any node in the first class is connected with every node in the second class, can be obtained by successive node duplications from the graph $K_{1,1}$ consisting of two nodes connected by an edge.

$K_{2,3}$ is generated by three vertex duplications from $K_{1,1}$ (2.2.62)

Therefore, we can deduce the spectrum of any such graph $K_{m,n}$: it has the eigenvalues 0 and 2 with multiplicity 1 each, and the eigenvalue 1 with multiplicity $m + n - 2$. Conversely, any graph with this spectrum is a complete bipartite graph $K_{m,n}$. In particular, all the graphs $K_{m,n}$ with the same value of $m + n$ of nodes have the same spectrum, that is, they are *isospectral*. In particular, the spectrum of Δ does not completely determine a graph. We also observe the following fact. For a complete graph K_N, for $N \to \infty$, all the eigenvalues converge to 1, see (2.2.44), except for $\lambda_0 = 0$. Therefore, in this limit, the difference between the spectra of a complete graph K_N and a complete bipartite graph $K_{m,n}$ with $m + n = N$ is only reflected by a single eigenvalue, the highest eigenvalue λ_K which remains 2 for $K_{m,n}$, but goes to 1 for K_N. To appreciate this phenomenon, we observe that K_N is the graph with the maximal number of 3-cycles, i.e., triangles, because every edge is contained in $N - 2$ triangles and every vertex is a vertex of $\binom{N-1}{2}$ triangles. $K_{m,n}$ does not contain any triangles, but otherwise is the graph with the maximal number of 4-cycles, in the sense that every vertex of the second class, that with n vertices, is a vertex of $\binom{m}{2}(n-1)$ 4-cycles, and analogously for the first class. From this observation, we also see that different such $K_{m,n}$ with the same sum $m+n$ are distinguished by their numbers of cycles. Therefore, this number, together with the spectrum, can uniquely identify a graph $K_{m,n}$.

These issues are further developed in [7, 8], and biological applications are given in [6, 9].

We now return to the issue of decomposing a graph by cutting edges. There exists an important relationship of this issue with the first eigenvalue λ_1 which we shall now describe. This is based on a quantity that is analogous to one introduced by Cheeger in Riemannian geometry, but had already been considered earlier in graph theory by Polya. We therefore call it the Polya-Cheeger constant. Letting $|E|$ denote the number of edges contained in an edge set E, the Polya-Cheeger constant is

$$h(\Gamma) := \inf_{E_0} \left\{ \frac{|E_0|}{\min(\sum_{i \in V_1} b_i, \sum_{i \in V_2} b_i)} \right\} \qquad (2.2.63)$$

where removing E_0 disconnects Γ into the components V_1, V_2. Thus, we try to break the graph up into two large components by removing only few edges. We may then repeat the process within those components to break them further up until we are no longer able to realize a small value of h.

We now derive elementary estimates for λ_1 from above and below in terms of the constant $h(\Gamma)$. Our reference here is [28] (that monograph also contains many other spectral estimates for graphs, as well as the original references; the analogy between the Cheeger estimate in Riemannian geometry and in graph theory was discovered in [35]). We start with the estimate from above and use the variational characterization (2.2.47). Let the edge set E_0 divide the graph into the two disjoint sets V_1, V_2 of nodes, and let V_1 be that with the smaller vertex sum $\sum b_i$. We consider a function v that is $=1$ on all the nodes in V_1 and $= -\alpha$ for some positive α on V_2. α is chosen so that the normalization $\sum_V b_i v(i) = 0$ holds, that is, $\sum_{i \in V_1} b_i - \sum_{i \in V_2} b_i \alpha = 0$. Since V_2 is the subset with the larger $\sum b_i$, we have $\alpha \leq 1$. Thus, for our choice of v, the quotient in (2.2.47) becomes $\leq \frac{(1+\alpha)^2 |E_0|}{\sum_{i \in V_1} b_i + \sum_{i \in V_2} b_i \alpha^2} = \frac{(\alpha+1)|E_0|}{\sum_{V_1} b_i} \leq 2 \frac{|E_0|}{\sum_{V_1} b_i}$. Since this holds for all such splittings of our graph Γ, we obtain from (2.2.63) and (2.2.47)

$$\lambda_1 \leq 2h(\Gamma). \qquad (2.2.64)$$

The estimate from below is slightly more subtle, and the estimate presented here works only for the choice

$$b_i = n_i. \qquad (2.2.65)$$

We consider the first eigenfunction u_1. Like all functions on our graph, we consider it to be defined on the nodes. We then interpolate it linearly (or monotonically) on the edges of Γ. Since u_1 is orthogonal to the constants (recall $\sum_i n_i u(i) = 0$), it has to change sign, and the zero set of our extension then divides Γ into two parts

Γ' and Γ'' by sign. W.l.o.g., Γ' is the part with fewer nodes. The points where (the extension of) $u_1 = 0$ are called boundary points. We now consider any function φ that is linear on the edges, 0 on the boundary, and positive elsewhere on the nodes and edges of Γ'. We also put $h'(\Gamma') := \inf_{E_1}\{\frac{|E_1|}{\sum_{i \in \Omega} n_i}\}$ where removing the edges in E_1 cuts out a subset Ω of the vertex set of Γ' that is disjoint from the boundary. We also use the identity

$$\sum_{e=(i,j)} |\varphi(i) - \varphi(j)| = \int_\sigma \sharp_e(\varphi = \sigma) d\sigma \tag{2.2.66}$$

where $\sharp_e(\varphi = \sigma)$ denotes the number of edges on which φ attains the value σ. This is illustrated by the following figure for the case of a cyclic graph with 5 vertices

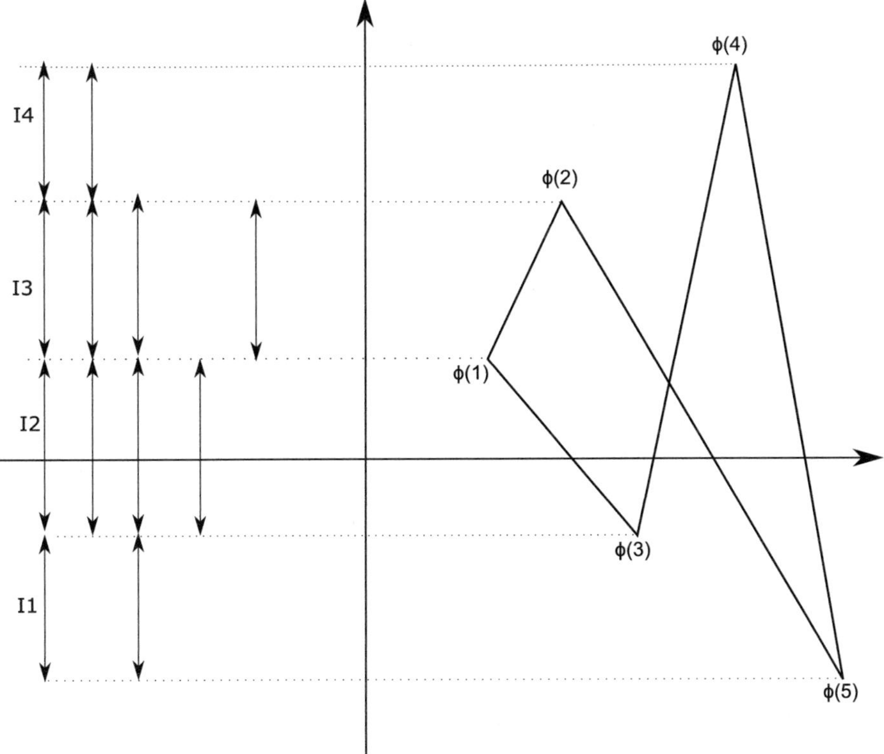

From (2.2.66), we proceed to

$$\sum_{e=(i,j)} |\varphi(i) - \varphi(j)| = \int_\sigma \frac{\sharp_e(\varphi = \sigma)}{\sum_{i:\varphi(i)\geq\sigma} n_i} \sum_{i:\varphi(i)\geq\sigma} n_i \, d\sigma$$

$$\geq \inf_\sigma \frac{\sharp_e(\varphi = \sigma)}{\sum_{i:\varphi(i)\geq\sigma} n_i} \int_s \sum_{i:\varphi(i)\geq s} n_i \, ds$$

$$= \inf_{\sigma} \frac{\sharp_e(\varphi = \sigma)}{\sum_{i:\varphi(i)\geq\sigma} n_i} \sum_i n_i |\varphi(i)|$$

$$\geq h'(\Gamma') \sum_i n_i |\varphi(i)|$$

when the sets $\varphi = \sigma$ and $\varphi \geq \sigma$ satisfy the conditions in the definition of $h'(\Gamma)$; that is, the infimum has to be taken over those $\sigma < \max \varphi$. Applying this to $\varphi = v^2$ for some function v on Γ' that vanishes on the boundary, we obtain

$$h(\Gamma') \sum_i n_i |v(i)|^2 \leq \sum_{e=(i,j)} |v(i)^2 - v(j)^2|$$

$$\leq \sum_{e=(i,j)} (|v(i)| + |v(j)|)|v(i) - v(j)|$$

$$\leq \sqrt{2}(\sum_i n_i |v(i)|^2)^{1/2}(\sum_{e=(i,j)} |v(i) - v(j)|^2)^{1/2}$$

from which

$$\frac{1}{2} h(\Gamma')^2 \sum_i n_i |v(i)|^2 \leq \sum_{e=(i,j)} |v(i) - v(j)|^2. \tag{2.2.67}$$

We now apply this to $v = u_1$, the first eigenfunction of our graph Γ. We have $h'(\Gamma') \geq h(\Gamma)$, since Γ' is the component with fewer nodes. We also have[7]

$$\lambda_1 \sum_{i\in\Gamma'} n_i u_1(i)^2 = \frac{1}{2} \sum_{i\in\Gamma'} \sum_{j\sim i} (u_1(i) - u_1(j))^2, \tag{2.2.68}$$

cf. (2.2.24) (this relation holds on both Γ' and Γ'' because u_1 vanishes on their common boundary).[8] Equation (2.2.67) and (2.2.68) yield the desired estimate (under the assumption (2.2.66))

$$\lambda_1 \geq \frac{1}{2} h(\Gamma)^2. \tag{2.2.69}$$

From (2.2.64) and (2.2.69), we also observe the inequality

$$h(\Gamma) \leq 4 \tag{2.2.70}$$

---

[7] We obtain the factor $1/2$ because we are now summing over vertices so that each edge gets counted twice.

[8] To see this, one adds nodes at the points where the edges have been cut, and extends functions by 0 on those nodes. These extended functions then satisfy the analogue of (2.2.19) on either part, as one sees by looking at the derivation of that relation and using the fact that the functions under consideration vanish at those new "boundary" nodes.

for any connected graph, when the weights b_i are the vertex degrees n_i. In fact, we can obtain a better estimate from (2.2.69). Since, as noted above in (2.2.44), (2.2.48), we always have $\lambda_1 \leq \frac{N}{N-1}$, we see directly that

$$h(\Gamma) \leq \sqrt{\frac{2N}{N-1}}. \tag{2.2.71}$$

Also, unless the graph is complete, we have $\lambda_1 \leq 1$, see (2.2.48), and therefore, for non-complete graphs, we have the estimate

$$h(\Gamma) \leq \sqrt{2}. \tag{2.2.72}$$

One can also think about the decomposition of a graph by removing vertices instead of edges. This issue is amenable to a similar treatment, and one can define a quantity analogous to $h(\Gamma)$ that has the number of vertices whose elimination is needed to disconnect the graph in the numerator; see [28] for details. Moreover, we can also define a dual Cheeger constant that can be utilized to control the largest eigenvalue, see [15].

The spectrum of the graph Laplacian is a useful tool to analyze biological networks, see [6, 8, 9, 10]. Whereas most computational problems on graphs are NP-hard or even NP-complete, and hence require a number of steps that grows exponentially with the number of vertices, the computation of the spectrum proceeds by linear algebra. Therefore, there exist algorithms that grow only like a polynomial of low order in the number of vertices. With current methods, one can determine the spectrum of graphs with about half a million nodes, and if one exploits the particular structures that can typically be found in empirical networks, one can handle even larger ones.

Therefore, one computes spectra of graphs in order to compare or distinguish biological and other networks by their spectral properties. We should remark at this point that the spectrum of its Laplacian does not always determine a graph uniquely. For instance, all complete bipartite graphs $K_{m,n}$ with the same total number $m + n$ of vertices have the same spectrum, i.e., they are isospectral, as we have already observed in the discussion after (2.2.61), with their spectrum consisting of 0 and 2 with multiplicity 1 each, and the eigenvalue 1 with multiplicity $m + n - 2$.

Nevertheless, as we have seen above, the spectrum reflects many important structural properties of a graph, like its decomposability.

Networks from specific domains, for instance protein-protein interaction networks (see e.g. [6]) usually share specific properties that distinguish them from networks from other domains. By investigating these specific spectral properties, one can then gain insight about the structure of such networks. For instance, a high multiplicity of the eigenvalue 1 in molecular networks may indicate gene duplications underlying the evolutionary history of such networks.

Many biological networks are, in fact, directed. In metabolic networks, there is the distinction between inputs and outputs of reactions, and one is interested in flows through such a directed network. In neuronal networks, information is transmitted from a presynaptic to a postsynaptic neuron, and one wants to understand the resulting dynamics. In food webs, trophic interactions ("who eats whom") are naturally unsymmetric. Therefore, the spectral theory of directed graphs has been systematically developed by F.Bauer, see [14]. Applications to biological networks are explored in [11].

2.3 Descendence Relations

2.3.1 Trees and Phylogenies

Trees are the formal tool for representing ancestor-descendent relations in biology and other fields. At first sight, the concept of a tree as defined below seems not appropriate for that task, however, when one thinks of parent-offspring relationships in sexually recombining species. There, the relationship graph, the so-called pedigree is branching in the backward direction because each individual has two parents, as well as in the forward direction because individuals on average have more than one offspring if the population is not going extinct. When one considers asexual reproduction, however, the situation becomes simpler because each individual then has only one parent, and branching can occur only forward in time when one considers the descendents over the generations of a single ancestor. This, perhaps, is not such an exciting problem, and, in fact, biologists are rather interested in trees for describing phylogenetic relationships between species instead of individuals. The endpoints of a tree, the so-called leaves (see below for the formal definitions), then correspond to a collection of recent species, and one tries to construct a tree in which the internal vertices represent ancestral species that are the common ancestors of all the species below them. Here, one usually assumes that speciation events are binary branchings, that is, one species splits into two daughter species. (In order to make this consistent, at least some biological taxonomists, the cladists, adopt the convention that whenever a new species branches off from an existing one, the remaining part of the latter then is also classified as a new species.) Traditionally, the similarities between species were gauged on the basis of morphological features, and paleontologists tried to identify the hypothetical ancestral species with ones documented in the fossil record. (In practice, this encounters many problems, but that is not our concern here.) Today, there exists a powerful alternative to that classical method, the comparison on the basis of genetic data. The idea is obvious, to take DNA samples from members of different species and count the differences so as to determine the genetic distances between the species. On the basis of those distances, a hierarchical grouping should be possible that can be represented by a tree. Of course, in practice, this is not so simple. First of all, the genetic samples need to be comparable. For that, one needs to identify DNA segments in the species representatives that are homologous to each other, that is, derived from the same ancestral sequence through a process of accumulation of mutations. Since besides point mutations in the DNA, there can also occur rearrangements like insertions, deletions, inversions, first the problem of sequence alignment needs to be addressed and solved for the samples at hand. This is usually done with the BLAST algorithm [2]. Next, one assumes that mutations occurred at the same rate in the different lineages, the hypothesis of the molecular clock. Otherwise, the number of genetic differences would not be a uniform measure of the time since branching from a common ancestor. Moreover, one needs to find genetic regions that have not been under selective pressure, but rather where there is a uniform probability of the retention of any mutation. Under stabilizing selection, most mutations are eliminated, and this would lead to an underestimate for the time since branching. For directed selection, in contrast, adaptive pressure leads to a more rapid accumulation of mutations and then to an overestimate of the time since branching.

Even if one can align the sequences successfully and eliminate selection effects, there still remain substantial problems. Often, the genetic distances vary with the genomic regions considered. Thus, depending on the DNA region considered, one might get a different tree. In that case, one might try to find some kind of compromise tree. That will depend on the criterion adopted, however, as we shall discuss a little more below. Sometimes, the data even do not fit into a tree because distances on a tree need to satisfy some necessary conditions discussed below. The question then is what substitute to choose for a tree, an issue that we shall also address below. Also, a species is not entirely homogeneous, and there are also genetic differences between the members of the same species (otherwise, evolution could not work by differential selection). Therefore, one needs to gauge intraspecies differences against interspecies ones. Finally, speciation is not an event that takes place at one clearly identifiable point in time, but rather is a gradual process of the accumulation of differences between different populations until reproductive barriers emerge that prevent further genetic mixing between those populations. Here, we need to invoke the species concept of modern biology. A species is defined as a population of organisms that can sexually produce viable and fertile offspring among them. In practice, however, sometimes that relationship is not necessarily transitive. That is, there can exist subpopulations $A_1, \ldots, A_k$ such that individuals from A_i can reproduce with those of A_{i+1} for all i, but those from A_1 are no longer able to reproduce with those from A_k. An example are the races of domestic dogs that range from rather large to very small ones. More generally, for the assembly of phylogenetic trees, species are considered as static ensembles, while in reality speciation is a temporally extended dynamic process inside groups of indidivuals (see the discussion in [20]). (As an aside, some of those population dynamics can be reconstructed on the basis of a statistical analysis of the distribution of alleles in recent populations, in particular from their deviations from equilibria defined by independence hypotheses.)

In spite of all these problems, phylogenetic tree reconstructions are a useful tool for many biologists. There is one issue, however, that calls for a generalization of the representation of phylogenies by trees. As L. Margulis emphasized, many genetic changes are not caused by mutations in inherited genomes, but rather by horizontal gene transfer through virusses and other processes [88]. That, of course, cannot be represented in a tree. Therefore, the tree formalism has recently been extended in

[112] to allow for horizontal gene transfer. On the other hand, over the course of evolution, organisms seem to have developed some protective mechanisms against such horizontal gene insertions, and the relative efficiency of those provides some justification for attempting to represent genetic data in a tree. In the light of all the difficulties mentioned above, it is then necessary to develop methods for finding trees that contain as few hypotheses as possible not supported by the available data.

We now start with the mathematical formalism as pioneered by Andreas Dress; we treat a particular class of graphs, the so-called trees. Our basic references are [107] and [37]. We shall not provide the proofs of the mathematical results discussed, but rather refer the reader to the literature.

We recall that a **tree** $T = (V, E)$ is a graph without cycles.

Lemma 2.3.1. *For a graph* $\Gamma = (V, E)$, *the following statements are equivalent:*

1. *Γ is a tree, that is, has no cycles.*
2. *For any two distinct vertices i, j, there exists a unique path of distinct vertices joining them (we shall call that path a "shortest path" even though we do not yet have specified a metric at this point—it will, however, turn out to be a shortest path for any metric on the tree).*
3. *$|V| = |E| + 1$.*
4. *The deletion of any edge disconnects Γ.*

The *proof* of this lemma is an easy exercise. Since for any graph $\Gamma = (V, E)$, we have $|V| \leq |E| + 1$, a tree thus is a graph with the minimal number of edges needed to connect a vertex set V.

The vertices of a tree that have degree 1 are called leaves. The other vertices are called interior vertices. Sometimes, it is convenient to exclude vertices of degree 2. A rooted tree is a tree with one distinguished vertex i_0, the root.

Rooted trees are the formal tool to represent hierarchical relationships between individual entities. We say that the vertex i_1 is above the vertex i_2, or in the phylogenetic interpretation to follow that i_1 is an ancestor of i_2, and i_2 a descendent of i_1, when the shortest path from i_0 to i_2 passes through i_1.

In phylogenies, the aim is the comparison between extant species. Those species then are represented as the leaves of some tree, and the rest of the tree then is built with the purpose that the interior vertices represent common ancestors of all those below some. Thus, the interior vertices may correspond to hypothetical species on which no data need to be available. Of course, paleontologists try to identify those interior vertices with fossil species, but the modern data usually consist of genetic data like pieces of DNA sequences for which one rarely has fossil samples. Thus, in paleontology, it is natural to allow for degree 2 vertices, representing ancestors of a single extant species that are documented in the fossil record. In molecular sequence analysis, however, one would exclude degree 2 vertices because all interior vertices represent hypothetical reconstructions of common ancestors of several descendent species.

In order to proceed with this formalization, we consider X-trees where X is some set. In applications, X of course is a or the data set. An X-tree is a tree $T = (V, E)$ together with a map $\phi : X \to V$ whose image contains all vertices of degrees 1 and 2. (In the rooted case, we do not require that the root be in the image of ϕ even though it may have degree ≤ 2.) The map need not be injective. For a phylogenetic (X-) tree, however, we require that ϕ be a bijection onto the leaves of T. In particular, such a phylogenetic tree has no vertices of degree 2. When every interior vertex has degree 3, we speak of a binary phylogenetic tree. This is a natural assumption in biology because, in evolution, a species can split into two daughter species, and each of those can then split again, and so on, but one does not see the emergence of three or more daughter species at the same time. In fact, much of phylogenetic tree reconstruction is about resolving the question in which temporal order the various splits into daughter species took place.

An X-split $A|B$ is a partition of X into two non-empty subsets A, B.[9] Thus, in biological applications, A might represent those members of X where a certain feature is present, and B those where that feature is absent.—Two such splits $A_1|B_1$ and $A_2|B_2$ are called compatible when at least one of the intersections $A_1 \cap A_2$, $A_1 \cap B_2$, $B_1 \cap A_2$, $B_1 \cap B_2$ is empty. If, say, $A_1 \cap B_2 = \emptyset$ then $A_1 \subset A_2$ and $B_2 \subset B_1$, and vice versa, and so, there is an alternative way of expressing compatibility of splits. When we have an X-tree (T, ϕ), then every edge e of T induces an X-split because it decomposes T into two subgraphs T_1, T_2 (which might include the degenerate case where one of them consists of a single vertex and no edges), and their preimages under ϕ then constitute a split of X. When we assume that the tree has no vertices of degree 2—which we shall henceforth do—different edges lead to subgraphs with different leaf sets, and therefore different edges induce different splits of X. Those splits then are compatible. We denote the splits of X induced by the X-tree (T, ϕ) by $\Sigma(T, \phi)$, or simply by $\Sigma(T)$ when the map ϕ is implicitly understood.

The converse question of what classes of splits of X come from X-trees is answered by the following result of Buneman

Theorem 2.3.1. *Given a collection Σ of X-splits, there exists an X-tree (T, ϕ) (which then is unique—up to isomorphism, of course) for which $\Sigma = \Sigma(T, \phi)$ precisely if all the splits in Σ are pairwise compatible.*

A tree carries an obvious metric, in the sense that we can quantify the distance between vertices i_1 and i_2 by counting the number of edges in the shortest path between them. More generally, we can assign positive weights $w(e)$ to the edges e and then take the sum of the weights of the edges in such a path as the distance $d(i_1, i_2)$.

When we consider a set X, there may already exist some distance function on X, and the question then emerges whether that distance is compatible with the metric on some X-tree. The answer is pretty simple, and in fact, we can even take something more general than a metric on X, namely a so-called dissimilarity map, that is, a non-negative map $\delta : X \times X \to \mathbb{R}$ with $\delta(x, x) = 0$ and otherwise positive, and

---

[9] That A and B yield a partition of X means that $A \cup B = X$ and $A \cap B = \emptyset$.

$\delta(x, y) = \delta(y, x)$ for all $x, y \in X$. For example, $\delta(x, y)$ could just count in how many characters (see below for a formal definition) the elements x and y differ. The question then is whether we can find an X-tree (T, ϕ) with weights $w(e)$ on its edges and associated distance function $d(., .)$ such that

$$\delta(x, y) = d(\phi(x), \phi(y)) \tag{2.3.1}$$

for all $x, y \in X$. In that case, we call δ a tree metric. The answer is

Theorem 2.3.2. *A dissimilarity map δ on X is a tree metric precisely if it satisfies the 4-point condition*

$$\delta(x, y) + \delta(z, w) \leq \max(\delta(x, z) + \delta(y, w), \delta(x, w) + \delta(y, z)) \tag{2.3.2}$$

for all $x, y, z, w \in X$.

In the sequel, (2.3.2) will give rise to two different issues. One is whether it holds or not for all points, and this issue is exemplified in the case where δ is the metric coming from a quadrilateral graph where x, w, y, z are arranged in cyclic order, for example $\delta(x, w) = \delta(w, y) = \delta(y, z) = \delta(z, x) = 1$ and $\delta(x, y) = \delta(z, w) = 2$. Thus, (2.3.2) is not satisfied here. The other issue arises when (2.3.2) is satisfied for all quadruples and consists in the question under which conditions we have even strict inequality for certain quadruples.

Since every edge e of an X-tree corresponds to a split σ of X, we can write a tree metric as

$$d = \sum_{\sigma \in \Sigma(T)} w(e_\sigma)\delta_\sigma \tag{2.3.3}$$

where e_σ is the edge inducing the split σ and

$$\delta_\sigma(i, j) = \begin{cases} 1 \text{ if } i, j \text{ are in different components of } T - e \\ 0 \text{ otherwise.} \end{cases}$$

The point here is that the edges e_σ occurring for $d(x, y)$ in (2.3.3) with $\delta_\sigma(x, y) = 1$ are precisely those contained in the shortest path from x to y.

This will now lead us to the decomposition theorem of Bandelt and Dress [12]. Let δ be a dissimilarity map on X. For a split $\sigma = A|B$ of X, we consider

$$i_\delta(\sigma) := \frac{1}{2} \min_{a_1, a_2 \in A, b_1, b_2 \in B} (\max(\delta(a_1, b_1) + \delta(a_2, b_2), \delta(a_1, b_2) + \delta(a_2, b_1))$$
$$- (\delta(a_1, a_2) + \delta(b_1, b_2))). \tag{2.3.4}$$

It is not required that the points a_1 and a_2 or b_1 and b_2 be different. For example, this expression can become negative when δ does not satisfy the triangle inequality:

take $a_1 = a_2 =: a$ and b_1, b_2 with $\delta(b_1, b_2) > \delta(b_1, a) + \delta(a, b_2)$.—In order to understand the significance of $i_\delta(\sigma)$ better, we consider some examples. These examples will be graphically displayed in the figure below. We first take a space $X = \{x, y, z, w\}$ consisting of 4 points, with the condition

$$\delta(x, y) + \delta(z, w) < \max(\delta(x, z) + \delta(y, w), \delta(x, w) + \delta(y, z)). \qquad (2.3.5)$$

If $\delta(x, y) = \delta(z, w) = 2$, $\delta(x, z) = \delta(x, w) = \delta(y, z) = \delta(y, w) = 3$, the split $\{x, y\}|\{z, w\}$ has index $i_\delta = 1/2$ and is induced from a tree with leaves x, y, z, w and interior nodes ξ, ζ with $\delta(x, \xi) = \delta(y, \xi) = \delta(\xi, \zeta) = \delta(z, \zeta) = \delta(w, \zeta) = 1$, see the following figure

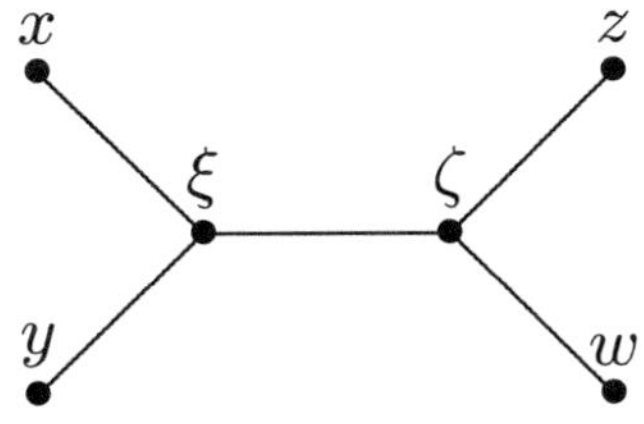

$$(2.3.6)$$

The splits $\{x, z\}|\{y, w\}$ or $\{x, w\}|\{y, z\}$, however, have $i_\delta(\sigma) = 0$ and are not induced by that tree metric. When we have equality in (2.3.5), say, $\delta(x, y) = \delta(z, w) = 2$, $\delta(x, z) = \delta(z, w) = \delta(y, z) = \delta(y, w) = 2$, the metric can still be represented by a tree metric, this time with a single interior vertex ξ that has distance 1 from all leaves. Thus, we have a 4-star

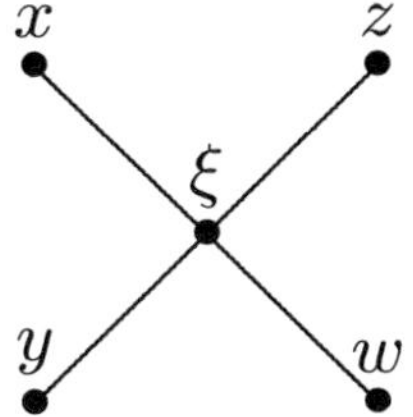

$$(2.3.7)$$

Here, there is no longer a natural grouping of the vertices into two pairs.—When instead $\delta(x, y) = \delta(z, w) = \delta(x, z) = \delta(y, w) = 3$, and $\delta(x, w) = \delta(y, z) = 4$, then (2.3.5) holds again. This time, we can represent the metric by a graph with 4 interior vertices ξ, η, ζ, ω that is not a tree. ξ, η, ζ, ω form a rectangle with $\delta(\xi, \eta) = \delta(\xi, \zeta) = \delta(\omega, \zeta) = \delta(\eta, \omega) = 1$, the other nontrivial distances between them being equal to 2, and with x connected to ξ, y to η, z to ζ, w to ω, all with distance 1. Thus, we need to insert an interior rectangle in order to represent the metric on a graph,

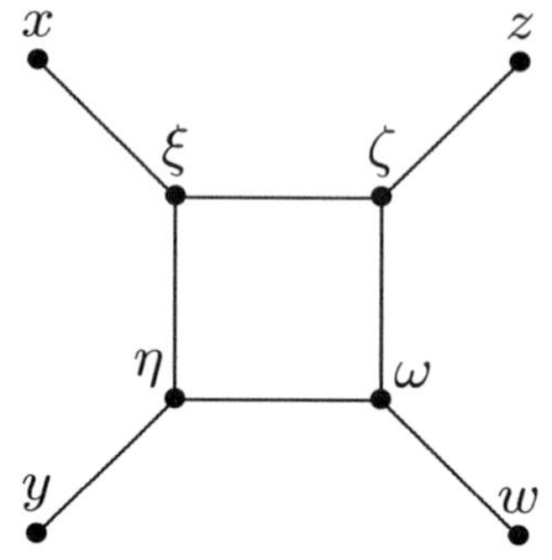

$$(2.3.8)$$

That rectangle then expresses the ambiguity in the dissimilarity map for a hierarchical grouping. Of course, the rectangle is in fact a square, and so there is some special symmetry. We therefore also consider the case where $\delta(x, y) = \delta(z, w) = 3$, $\delta(x, z) = \delta(y, w) = 4$, $\delta(x, w) = \delta(y, z) = 5$. In that case, we again insert 4 interior vertices ξ, η, ζ, ω that form a rectangle, this time with $\delta(\xi, \eta) = \delta(\zeta, \omega) = 1$, $\delta(\xi, \zeta) = \delta(\omega, \eta) = 2$,

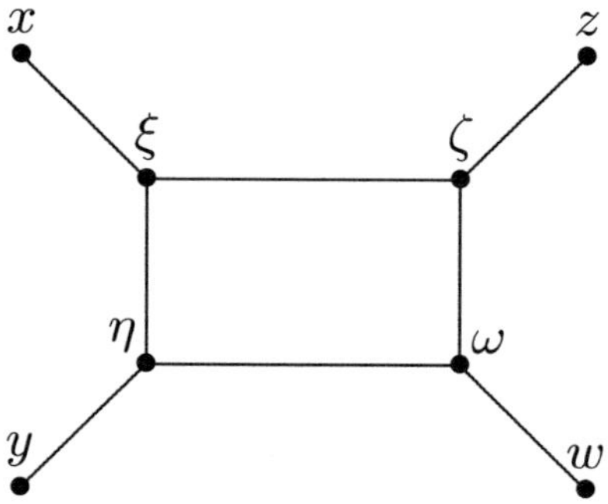

$$(2.3.9)$$

In any case, when we have such a rectangle, we produce splits by cutting pairs of parallel edges. Cutting the edges between ξ and ζ and between η and ω, for example, produces the split $\{x, y\}|\{z, w\}$. Cutting the edges between ξ and η and between ζ and ω instead produces the split $\{x, z\}|\{y, w\}$. Now, in contrast to the tree case, both these splits have $i_\delta(\sigma) > 0$. The split $\{x, w\}|\{y, z\}$, however, has $i_\delta(\sigma) < 0$. In the tree case, interchanging x with y or z with w would not have made any difference for the distances between those 4 vertices,

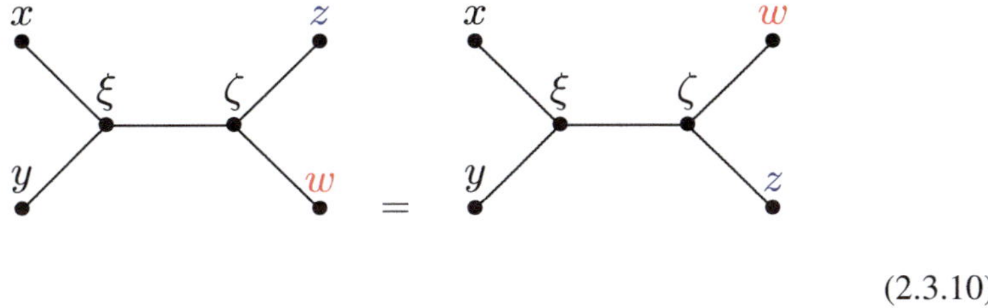

$$(2.3.10)$$

but this is no longer so in the rectangle case.

After this example, let us return to the general case. When δ is a tree metric from an X-tree (T, ϕ), the split σ of X is induced by that X-tree precisely if $i_\delta(\sigma) > 0$. In that case, we then have $i_\delta(\sigma) = w(e_\sigma)$ for the weight of the edge inducing the split. And we can rewrite (2.3.3) then as

$$\delta = \sum_{\sigma \ X\text{-split with } i_\delta(\sigma)>0} i_\delta(\sigma)\delta_\sigma. \qquad (2.3.11)$$

The split decomposition theorem of Bandelt and Dress [12] then says that every dissimilarity map can be written as a sum over such tree metrics plus a remainder that has no splits with $i_\delta(\sigma) > 0$:

Theorem 2.3.3. *Let δ be any dissimilarity map on X. We then have a decomposition*

$$\delta = \delta_0 + \sum_{\sigma \ X\text{-split with } i_\delta(\sigma)>0} i_\delta(\sigma)\delta_\sigma \qquad (2.3.12)$$

where δ_0 admits no splits with $i_\delta(\sigma) > 0$.

The star in our above example (2.3.7) admits no splits into pairs of points with $i_\delta(\sigma) > 0$. This is an undesirable situation in phylogenetic tree reconstruction because the grouping of the four vertices into pairs is ambiguous. However, when we split off a single point from the remaining three, we get $i_\delta(\sigma) > 0$. The simplest example of a metric space admitting no splits at all with $i_\delta(\sigma) > 0$ is given by 5 points x, y, z, w, v with $d(x, v) = d(y, z) = d(z, w) = d(y, w) = 2$ and the other distances between different points all being one. To describe this metric space somewhat differently, we take the two sets $A := \{x, v\}$, $B := \{y, z, w\}$ and connect each point in A with every point in B by an edge of length 1. Thus, we see that the graph constructed in this way is the bipartite graph $K_{2,3}$ of (2.2.9).—For this example, then δ_0 is nontrivial, and moreover, $d = \delta_0$.

When, conversely, δ_0 vanishes, the dissimilarity map δ is called totally decomposable. We recall that in the above example with the interior rectangle, the splits $\{x, y\}|\{z, w\}$ and $\{x, z\}|\{y, w\}$ both have positive $i_\delta(\sigma)$, and they decompose the metric. Thus, a totally decomposable metric need not be a tree metric. The problem of this example for phylogenetic tree reconstruction is that there is no unique split that decomposes the dissimilarity map, that is, on the basis of the dissimilarity map, we do not know how to group the elements. Another, larger, example of this type, that is, of a totally decomposable metric that is not a tree metric and where therefore the groupings of the elements are not unique, is displayed in the next figure.

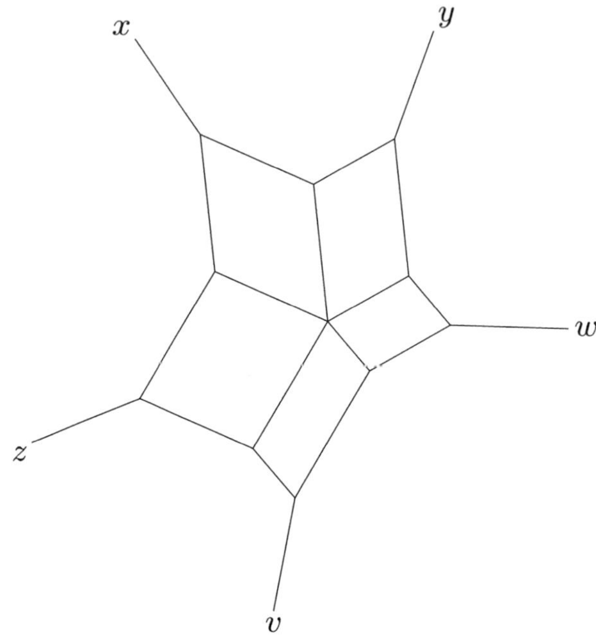

Bandelt and Dress[12] proved that a dissimilarity map δ is totally decomposable iff for all $x, y, z, v, w \in X$,

$$i_\delta(\{x, y\}|\{z, v\}) \leq i_\delta(\{x, w\}|\{z, v\}) + i_\delta(\{x, y\}|\{z, w\}). \tag{2.3.13}$$

Splits are decompositions of X into two subsets. More generally, we can consider characters, that is, functions $\chi : X_0 \to S$ where $\emptyset \neq X_0 \subset X$ and S is a finite set, the set of characters states. χ is called non-trivial if there are at least two character states that are each assumed by more than one element of X_0. We say that the character χ factors through the X-tree (T, ϕ) when there exists $\chi' : T \to S$ (here, we mean by a function on the tree T a function that is defined on the vertices of T) with $\chi = \chi' \circ \phi_{|X_0}$. Such a character χ that factors through the X-tree (T, ϕ) is called convex on T if for each $a \in S$, the subgraph with vertex set $(\chi')^{-1}(a)$ is connected. This is equivalent to the existence, for every pair a, b of different character states, of an X-split $A|B$ of T with $(\chi')^{-1}(a) \subset A$ and $(\chi')^{-1}(b) \subset B$.
The concept of character convexity is fundamental for the phylogenetic systematics developed by W.Hennig, the so-called cladism [58]. There, one wants to identify

monophyletic groups, that is rooted subtrees of phylogenetic trees that contain all the descendents of that vertex that is declared to be the common ancestor and made the root of the subtree. For example, in standard zoological systematics, vertebrates constitute a monophyletic group while fish don't because the other vertebrate groups (amphibians, reptiles, birds, and mammals) are also descendents of fish; in fact, here only birds and mammals are monophyletic in the sense of cladism. We consider an ancestral species A with daughter species A_1 and A_2.[10] Of course, this can be represented by a tree with root A and leaves A_1, A_2. Suppose that a character state a in A is preserved in A_1, but changed into a' in A_2. Now suppose that that species A_2 further splits into two daughter species A_{21} and A_{22}. We then get a new tree with leaves A_1, A_{21}, A_{22} if there are no further splittings while A_2 now is an interior node of degree 3. We consider two cases as displayed in the following figure.

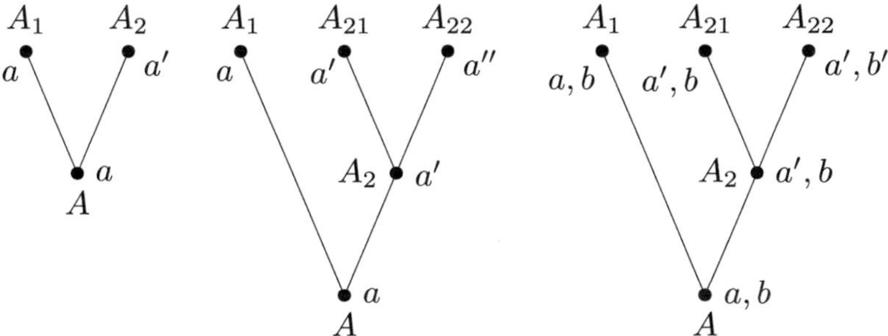

In the first case, A_{21} preserves the state a' while in A_{22} it is further transformed into a''. In the other case, both of them preserve a', but in A_{22} the state of some other character is transformed into the value b' from the common value b shared by A, A_1, A_2, A_{21}. In such a situation, the ancestral states a, b are called plesiomorph, the derived states a', a'', b' apomorph. These are relative concepts because a' is plesiomorph compared with a'', that is, when we only consider the subtree with root A_2 and leaves A_{21}, A_{22}. Two species sharing the same plesiomorph state of a character are called symplesiomorphic w.r.t. that character, those sharing an apomorphic state are called synapomorphic. In the last example, A_1 and A_{21} are symplesiomorphic for b while A_{21} and A_{22} are synapomorphic w.r.t. a'. In the preceding example, where A_{22} had the character state a'', the states a', a'' together constitute a synapomorphy between A_{21} and A_{22}. Only synapomorphy, but not symplesiomorphy, can be an indication of a monophyletic group. Here then enters the convexity assumption. Namely, in order to be able to use shared derived characters, that is, synapomorphies for identifying monophyletic groups, we must exclude the following two possibilities:

1. Reversion: In the last example, A_{22}, instead of assuming the new state a'', reverts to the ancestral state a.

---

[10] It is a basic principle of cladism that whenever a new species splits off from some line, the remaining part of that line is also classified as a new species. This makes the systematics amenable to tree representations. Moreover, from the morphological approach underlying cladism that is based on paleontological data, any two species differ in the state of at least one character.

2. Convergence: In the same example, A_1, instead of keeping the state b, assumes the same state b' that originated in the species A_{22} while A_{21} kept b.

Of course, there exist biological examples for either possibility. Snakes have lost the limbs that their ancestors had gained. Birds, bats, and insects have independently developed wings. In fact, the wings of birds and bats are plesiomorph when considered as forelimbs, but not as wings. Sometimes, the distinction between plesiomorphy and apomorphy is not clear or needs to be reconsidered in the light of genetic sequence data. For example, it had been thought for a long time that the eyes in arthropods, molluscs, and vertebrates are examples of a convergent evolution. It has been discovered, however, that eye formation in all these lineages is directed by the same master control gene, called *Pax6*, from the class of homeotic (Hox) genes [101]. An uncontroversial[11] example of convergence is mimikry where one species imitates the coloration or other pattern of an unrelated species that is avoided by predators. In any case, reversion and convergence are relatively rare in biological evolution, however. Both these possibilities are excluded by character convexity.

When one has several characters, one wants to find a single X-tree for which all of them are convex. When such a tree exists, these characters are called compatible. As for compatibility of splits, there exists a theorem characterizing the compatibility of characters, but since the formulation is more complicated we refer to [107].

When working with biological data, typically not all the characters are compatible, and one then wishes to quantify that non-compatibility and construct a tree that comes as close as possible to rendering all the characters convex. This is the idea of parsimony. More precisely, given a function f on the vertex set V of a graph Γ, the changing number of f is the number of those edges of Γ on whose endpoints f assumes different values, that is, the number of all edges $e = (i, j)$ with $f(i) \neq f(j)$. Let now $\chi : X_0 \to S$ be a character that factors through the X-tree (T, ϕ), with $\chi = \chi' \circ \phi_{|X_0}$ as above. Here, we are assuming that χ' is already defined on all the vertices of T. Of course, it is then arbitrary how to define χ' on those vertices of T that are not in the image of $\phi(X_0)$, in case ϕ is not surjective on X_0. For a character χ, we then define its parsimony score $s(\chi, T)$ for the X-tree (T, ϕ) as the minimal changing number of all those extensions χ' on T that factor χ. Given a set of characters, its parsimony score on an X-tree then is simply the sum of the individual parsimony scores. A maximal parsimony X-tree for that set of characters then is one that minimizes that parsimony score.

It is not difficult to see that the parsimony score is related to character convexity. In fact, given a character that assumes ν different states, the so-called homeoplasy of the character χ on T

$$h(\chi, T) := s(\chi, T) - \nu + 1 \geq 0 \tag{2.3.14}$$

---

[11] at least as long as one does not look at the underlying genetic mechanisms; in fact, it may well turn out in a given example that the imitation of a pattern is produced by the same kind of genetic regulatory mechanism as the imitated pattern, or at least the general framework of that genetic regulation might be derived from some common ancestor

with equality precisely if χ is convex on T. Thus, the total homeoplasy of a character set, the sum of the individual homeoplasies, is also non-negative and vanishes precisely when the characters are compatible.

The concept of maximum parsimony trees is not without difficulties, both conceptually and mathematically. The conceptual difficulties arise from the arbitrariness in the definition and choice of characters. It is a fundamental problem in paleontology and morphology to clearly state what a character is and to decide which characters are independent of each other. Of course, large sets of dependent characters would bias the parsimony concept. The mathematical problems become clear when one considers stochastic processes on trees and other graphs. One then realizes that any method of reconstructing a structure from a data set depends on a model for the underlying process that created the data.

A standard problem is to amalgate phylogenetic relationships between subsets of X as expressed in trees into an encompassing tree representing all of X. Of course, the issue of compatibility will arise again. The smallest meaningful subsets here consist of 4 elements and are called quartets, and trees with 4 leaves are called quartet trees. Also, if one has data about the relationships between the elements of X and wants to construct a tree or, more generally, find out whether these relationships fit into a tree, a natural strategy is to first construct all local quartet trees and then assemble those into a common tree. When we have a collection $\mathcal{Q}$ of quartet trees that contains exactly one quartet tree $\{a, b\}|\{c, d\}$ for every quartet $Y = \{a, b, c, d\}$ of X, then, as discovered by Colonius and Schulze [29], there exists a unique X-tree containing all these quartet tree iff the following two quartet rules hold for all $a, b, c, d, e \in X$:

1.
$$\text{If } \{a, b\}|\{c, d\}, \{a, b\}|\{d, e\} \in \mathcal{Q}, \text{ with } c \neq e, \text{ then } \{a, b\}|\{c, e\} \in \mathcal{Q}$$

2.
$$\text{If } \{a, b\}|\{c, d\}, \{a, c\}|\{d, e\} \in \mathcal{Q}, \text{ then } \{a, b\}|\{c, e\} \in \mathcal{Q}.$$

In practice, of course, these rules will be violated for some quintuples of elements of X, and one therefore cannot construct a tree.

There are other, in fact infinitely many, quartet rules. If $\mathcal{Q}$ does not contain a quartet tree for every quartet in X, that is, if we only have a subcollection of quartet trees, then we need to invoke more of those rules to check for compatibility, see [107] for more on this topic. For an algorithm for the (re)construction of a tree from quartets, see [114].

2.3.2 Genealogies (Pedigrees)

While species can be considered as important biological entities in their own right, the ancestor-descendent relationships in phylogenetic trees can also be viewed as accumulated genealogies of the individuals constituting the populations underlying the species. Thus, let us consider those genealogies a little, even though they in turn can be viewed as combinations of inheritance processes of genes passed on from parents to offspring. The latter, in fact, will lead us back to trees below.

The genealogy or pedigree of an individual in a sexually recombining population is a directed graph. Each individual has two incoming links from its parents while the number of outgoing links counts its offspring. Since no individual can be a descendent of its own offspring, or an ancestor of its own parents, the graph is acyclic (it has no directed cycles; the underlying undirected graph may well have cycles as the result of inbreeding in the population). The nodes without outgoing links represent those individuals that did not produce or have not yet produced offspring.

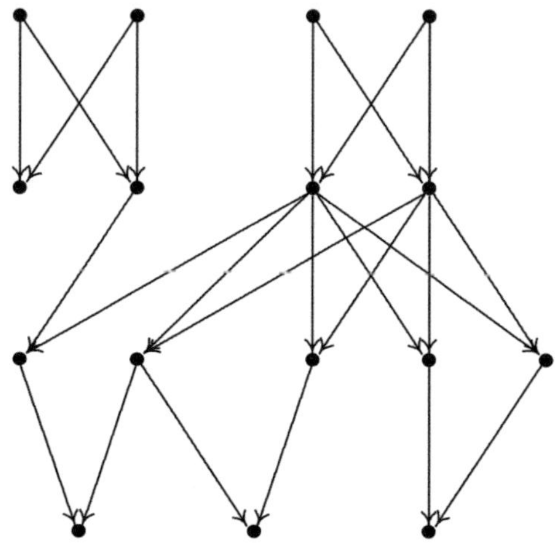

A pedigree graph; time runs downwards   (2.3.15)

In the pedigree graph (2.3.15), in the ancestral generation 1, we have two pairs that produce two offspring each. In generation 2, one individual leaves no offspring whereas another one contributes to five of them. In contrast, in generation 3, every individual leaves one or two descendents in generation 4. In a bisexual population, we can also identify two subgraphs, one corresponding to the females and the other to the males. In those subgraphs, every node then has precisely one incoming arrow. We shall return to this issue in Sect. 2.3.3.

When the graph represents a population history, one can essentialize it by pruning all the vertices without outgoing links that correspond to individuals having died without leaving offspring. This will then be an iterative process because in the next step one would have to prune those vertices that have outgoing links only to vertices that have been pruned in the previous step. In that manner, one iteratively eliminates all vertices that do not have living descendents. Thus, one is left with the ancestral relationships leading to the present population.

Since one does not want to extend the pedigree to the infinite past, one starts with some ancestral population. The essentialized pedigree then contains only those members of the ancestral population that have descendents in the present generation. If one moves further to the next generation, then some of those ancestors may cease to have descendents and therefore will get eliminated. Some of those ancestors, called the lucky ones, however, will turn out to be ancestors of all members of the present population, and they will therefore also leave descendents in all future generations, until the entire population goes extinct.

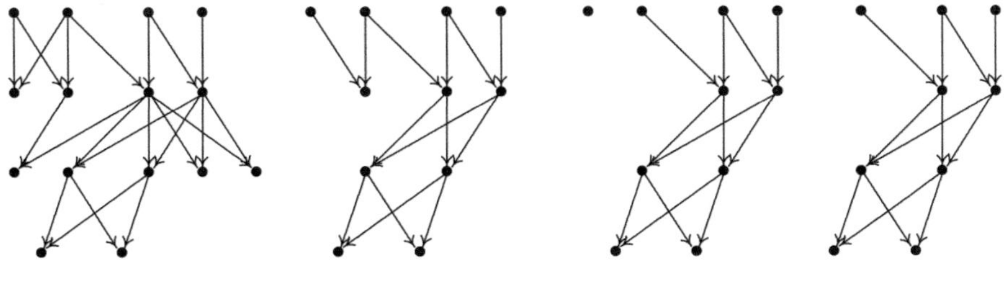

A pedigree graph and its prunings   (2.3.16)

In the pedigree graph of (2.3.16), there are three such lucky ancestors from whom the current population of two individual descends.

Often, one assumes that the different generations do not overlap, as in (2.3.15), (2.3.16). The generations can then be labelled by their distance from the ancestral one, and links always go from generation n to generation $n + 1$.

Also, from the pedigree graph, one can construct another graph expressing mating relationships. In that graph, there is an (undirected) edge between two individuals when they have produced offspring together. When the species is bisexual or dioecious, that is, has separate sexes, the mating graph is bipartite, the two classes corresponding to the females and the males. The mating graph usually is not connected, however, therefore, strictly speaking, violating our definition of a graph. When the population is strictly monogamous, the graph consists of disjoint pairs only, after we have essentialized it and eliminated all the bachelors and spinsters.

Of course, this is all rather simple. Later on, when we consider stochastic branching processes, pedigrees of sexually recombining populations become rather difficult, but for the moment we leave the subject and turn to

2.3.3 Gene Genealogies (Coalescents)

The pedigree just considered for a dioecious population, i.e., with two different sexes, contains two trees (more precisely, so called forests, that is, not necessarily connected unions of trees) as subgraphs, namely those corresponding to the male and the female individuals. Let us take one of them, say the female one. Thus, we only consider mother-daughter relationships. For two individuals, we can then ask when their lineages coalesce or merge back in the past, that is, how many generations back they had the same female ancestor. For two sisters, we need only go one generation back, as they have the same mother, while first (in the female line) cousins share a maternal grandmother, and so on. Once the lineages coalesce, they will stay together all the way back to the ancestral population. Of course, in principle, they may never merge, that is the two females under consideration may be descendents of different females in the ancestral population. When we go sufficiently many generations back into the past, however, with overwhelming probability, all presently living females in the populations will descend from the same ancestral female, the "Eve". All other females in that ancestral population will then have no descendents from an uninterrupted female line in the present populations; of course they may or may not have descendents from some lineages that include some males. As already described above, we can essentialize the graph by eliminating all females without female descendents in an iterative manner so that only those remain that have an uninterrupted line of female descendents down to the present sample. When we do coalescence theory, that is, follow the ancestry of the present sample back in time, then, in fact, those eliminated individuals will never occur in the consideration. This represents an enormous simplification in practice when compared with considering the forward branching process for the (female) descendents of an ancestral populations where all descendents will occur regardless of whether they contribute to future generations or not.

Let us consider this scenario in more detail in a simple example that will lead us to the Wright-Fisher model of population genetics. We consider a population with non-overlapping generations, and we assume that the size of the population remains constant $= 2N$ across generations. We also assume, for simplicity, that the sex ratio remains constant and equal so that we are dealing with a population of N females. The assumption of the Wright-Fisher model is that, given generation n, consisting of a population of N individuals, generation $n + 1$ is (mathematically) created by choosing N times randomly and independently an individual from generation n as mother.

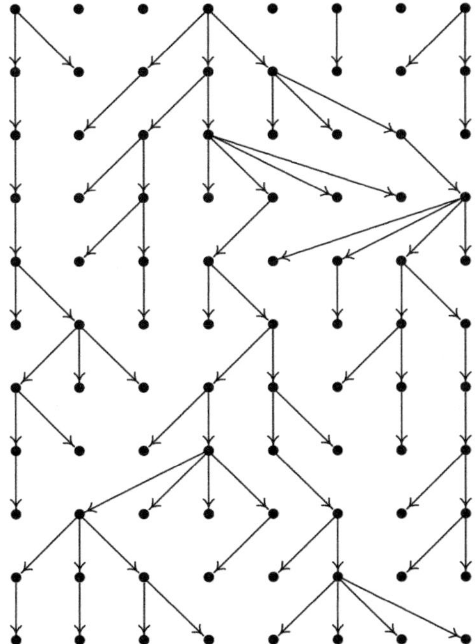

A Wright-Fisher genealogy with 8 individuals and 11 generations;   (2.3.17)

note that we have arranged the order of the individuals

in each generation so as to render the scheme clearer.

The 4th individual from generation 1 is the sole ancestors of all individuals in generation 10,

and the 5th individual from generation 6 is the sole ancestors of all individuals in generation 11.

Here, it is assumed that the population is entirely homogeneous, or, in more biological terms, that all members are equally fit, so that at each selection step, each member has the same chance of being chosen. Also, creating daughters does not affect the fitness, and so, the chance to be chosen at a given step does not depend on how often one has already been chosen in previous steps. Putting it another way, each individual in generation $n + 1$ picks individual j in generation n with probability $1/N$ as its mother, and this sampling is carried out N times with replacement. If d_j is the number of daughters of individual j, we thus have for the probability of having ν daughters

$$p(d_j = \nu) = \binom{N}{\nu} (\frac{1}{N})^{\nu} (1 - \frac{1}{N})^{N-\nu}. \qquad (2.3.18)$$

This is a binomial distribution, $Bi(N, \frac{1}{N})$, and so, the number of daughters of a given female is binomially distributed. The binomial distribution will be introduced more systematically in Chap. 3.1, see (3.1.8). The expectation value is

$$E(d_j) = N\frac{1}{N} = 1 \tag{2.3.19}$$

which of course reflects the fact that the population size is constant, and the variance is

$$Var(d_j) = N\frac{1}{N}(1 - \frac{1}{N}) = 1 - \frac{1}{N} \tag{2.3.20}$$

(see (3.1.20) below). The correlation between the numbers of daughters of different females j, k is

$$Cor(d_j, d_k) = \frac{Cov(d_j, d_k)}{\sqrt{Var(d_j)Var(d_k)}} = -\frac{1}{N-1}. \tag{2.3.21}$$

The correlation is negative, again because the population size is constant, and therefore, when j has many daughters, there is less room for k to have many daughters as well (when we already know that an individual different from j has one daughter, then the expected number of daughters of j is reduced to $\frac{N-1}{N}$ in place of the value 1 of (2.3.19)). This effect is rather small in large populations.
For large N, the binomial distribution $Bi(N, \frac{1}{N})$ is approximated by a Poisson distribution

$$p(d_j = \nu) \approx \frac{1}{\nu!}e^{-1} \tag{2.3.22}$$

with mean and variance $=1$ (see (3.1.8), (3.1.9) in Chap. 3.1 below). In particular, the probability of having no daughters is

$$p(d_j = 0) \approx e^{-1} \approx .37 \tag{2.3.23}$$

while then the probability to have at least one daughter becomes

$$p(d_j > 0) \approx 1 - e^{-1} \approx .63 \tag{2.3.24}$$

Therefore, the present population descends from a fraction of about $.63^n$ females n generations ago. Of course, this eventually goes to 0 for large n which leads to the absurd result that the present females derive from fewer than one individual in the ancestral generation. Of course, the puzzle is resolved by observing that these approximations were only valid for large population sizes. For small populations, a more refined analysis is needed. This is the subject of coalescence theory, originally founded by J. Kingman [81].

Again, we stay with our simple example and ask for the distribution of the number T_2 of generations that we need to go back in time to find a common ancestor of two individuals from the present population. That is, we seek the time to the most recent common ancestor (MRCA) of the two individuals. The probability that the two individuals have the same mother, that is, that the MRCA is found already in the first generation from the past, is $\frac{1}{N}$ because once we have identified the mother of the first individual, the probability that the second one has the same mother is $\frac{1}{N}$. Thus, the two have different mothers with probability $1 - \frac{1}{N}$. Iteratively, the chance to find the MRCA n generations back then is

$$p(T_2 = n) = (1 - \frac{1}{N})^{n-1}\frac{1}{N} \tag{2.3.25}$$

because they then have different ancestors in $n - 1$ generations. This is a geometric distribution, and its mean is

$$E(T_2) = \frac{1}{\frac{1}{N}} = N \tag{2.3.26}$$

which is equal to the population size.

In a similar manner, we can consider the time to find the MRCA for M individuals. The probability that m individuals have all different mothers is

$$\frac{N-1}{N}\frac{N-2}{N}\cdots\frac{N-m+1}{N} = \prod_{\mu=1}^{m-1}(1 - \frac{\mu}{N}) = 1 - \binom{m}{2}\frac{1}{N} + O(\frac{1}{N^2}) \tag{2.3.27}$$

because when the mother of the first individual is determined, there are $N - 1$ possibilities for the second to have a different mother, and when that is also determined, there remain $N - 2$ possibilities for the third individual to have a mother different from the previous two, and so on. Thus, neglecting terms of order $\frac{1}{N^2}$ for a large population size N, a coalescence event occurs in a given generation with probability $\binom{m}{2}\frac{1}{N}$, while no coalescence event occurs with probability $1 - \binom{m}{2}\frac{1}{N}$. Thus, the probability distribution for the time T_m of a coalescence event that reduces the number of different ancestors from m to $m - 1$

$$p(T_m = n) = (1 - \binom{m}{2}\frac{1}{N})^{n-1}\binom{m}{2}\frac{1}{N}. \tag{2.3.28}$$

In analogy to (2.3.26), we have

$$E(T_m) = \frac{1}{\binom{m}{2}\frac{1}{N}} = \frac{2N}{(m-1)m}. \tag{2.3.29}$$

When we then want to go back from M individuals to a single ancestor, we need to consider all the coalescent events from m to $m-1$ for $m = 2, \ldots, M$. Since the times for these events are independent of each other, the expected number of generations back in the past for M female individuals to have a single female ancestor is

$$\sum_{m=2}^{M} E(T_m) = \sum_{m=2}^{M} \frac{1}{\binom{m}{2}\frac{1}{N}} = \sum_{m=2}^{M} \frac{2N}{(m-1)m} = 2N(1 - \frac{1}{M}). \tag{2.3.30}$$

In other words, this is the expected height (measured in number of generations) of the tree starting with a single female ancestor and leading to the present ensemble of M females. When we compare (2.3.30) with (2.3.26), we see that the latter is less than 2 times the former. This means that the final step of reducing the number of ancestors from 2 to 1 typically takes at least half the time of the whole process. Thus, the long branches of the tree arise when there are only few females in the ancestry of the sample.

One can, of course, perform the same analysis with males in place of females. Let us insert a small variation, however, to account for the fact that in many animal species, like most mammals, and also in many human societies, the variance in the number of offspring for males is considerably higher than for females while obviously the expectation value is the same, assuming that the population is in gender equilibrium (that issue will be treated in Sect. 5.1). This higher variance is easy to achieve in our model. The simplest version just stipulates that in each generation only a certain fraction $0 < q < 1$ of the number of males is having offspring at all. When we then look for the father of an individual, each of those ones is taken with probability $1/qN$ while the other ones are simply discarded. Thus, two individuals now stand a chance of $1/qN$ of having the same father. Thus, N gets replaced by qN in all formulae. In particular, the expectation values for the waiting times in (2.3.26), (2.3.30) are shortened by a factor q, and we expect to find the MRCA in the male line correspondingly fewer generations ago than that in the female line. In other words, "Adam" lived many generations after "Eve". (In fact, it has recently been discovered that there is a small number of males of African origin that carry Y-chromosomes of different origin [90]. (The Y-chromosomes determine the male gender; a female possesses two X-chromosomes, a male one X- and one Y-chromosome; the latter are therefore passed on only the male line.) Thus "Adam" is not the male ancestor of all living humans, but only of the vast majority of them.)

Coalescence theory is mainly interested in describing the ancestry of genes, or more precisely, of DNA segments, rather than of individuals. Formally, the basic scenario is the same, however, and therefore, we have described the basic situation above for the more intuitive case of individuals. The basic scenario neglects the issues of mutation and recombination. In order to exclude recombination, there are two possibilities, one of significance for biological data, the other solely for modeling purposes. The first one consists in considering those DNA segments that do not recombine. One class is given by non-nuclear mitochondrial DNA that is only contained in egg cells, but not in sperm, and therefore is only passed on in the female line. This, in fact, makes the above example of female lineages relevant for treating biological data. The other example is the Y-chromosome in humans and other mammals which is only carried by males (and determines the male gender) and therefore is only transmitted in the male lineage. The mathematically convenient solution, in contrast, is to simply consider the smallest DNA segments, the single nucleotides. The biological problem here is that even though each nucleotide is derived from a unique parent, this usually cannot be identified from genetic data because in a given species, at most positions, most members share the same nucleotide. Nevertheless, for so-called SNPs, single nucleotide polymorphisms, consideration of single nucleotide positions can contain some useful population biological information. Even when we consider single nucleotide positions, however, we only get rid of the problem of recombination while absence of mutations, and of other processes of genetic rearrangement, then is still a hypothesis imposed. For simplicity of the model, we also consider the haploid case where each individual has only one set of genes. Thus, each such DNA segment in an individual is derived from one of the parents. (In the diploid case, each individual has two sets of genes. The genes corresponding to each other in those sets are selected from different parents, that is, one is taken from the mother and the complementary one from the father. This imposes additional restrictions when compared with the haploid case, but typically their effect is not so prominent.) For single nucleotides in the haploid case which we shall now consider the situation is formally the same as that where one of the two parents of each individual is its mother. So, one might call that individual that gives the nucleotide in question the nucleotide parent for that particular position in the DNA. When in turn we consider only nucleotide parents, for a fixed position, then the situation is the same as before, with the only formal difference that two siblings can derive the nucleotide at that position from different parents. Therefore, the size of the population that has to be taken into account is $2N$ in place of N.

In any case, for each such nucleotide position, we can perform the coalescent analysis and find the expected number of generations for having a single ancestor. Of course, the ancestors for the different positions will in general be different individuals. We can then also ask the following questions:

1. What is the expected number of generations for finding a single ancestor for each position? That is, what is the expected maximal height of the coalescence trees for a given population?
2. In the corresponding ancestral population, how many individuals are ancestors for some position for the present sample? Those lucky ancestors then are ancestors of every individual in the present sample whereas the remaining members of the ancestral population then are not genetically represented at all in that present sample because for each position, there is only one ancestor by assumption.
3. Actually, the last issue is a little more subtle. In principle, an individual in the ancestral population can be a genealogical ancestor of a present one without being genetically represented in the latter. What are the chances for that? Here, one should essentially use some tree counting arguments. The pedigree graph contains many trees with a root in the ancestral population and leading down

to the present population, and each such root then represents a genealogical ancestor. Not all of these trees, however, arise from the coalescence processes just investigated.

So far, only some partial answers are known to these questions, see [57] for a brief discussion and references. Here is an example that exhibits some of the problems. Until relatively recently, modern man, Homo sapiens sapiens, was not the only human species. The best known other such species are the Neanderthals who became extinct less than 30,000 years ago. This raises the question whether these different human species lived just alongside each other for a certain period, and perhaps violently competed, or whether they also mixed and interbred to a certain degree. The question is who are the ancestors of present humans, only some group of individuals that originated in Africa and whose descendents then spread to the other continents, or whether other human species that had lived in Africa and Eurasia prior to the spread of this lineage also contributed to our genomes. This question has been addressed in recent years both via the analysis of the genomes of living people from various populations in the world and from the analysis of the DNA of the bone fossils of Neanderthals and other human species, to the extent that this is technically feasible. First, it was found that in the mitochondrial DNA, no evidence of an admixture can be found. This DNA is not contained in the nucleus of a cell, but only in some organelles. It is therefore passed on only along the female line, as male sperm does not contain those organelles, but only female egg cells do. However, the more recent sequencing of large parts of the Neanderthal genome [54] showed that Eurasians do carry some percentage of DNA inherited from Neanderthals. Thus, the coalescence trees for different parts of the human genome are significantly different.

Exercises for This Chapter

1. This exercise introduces the perhaps best known combinatorial design problem. A Hadamard matrix is an $n \times n$ matrix whose entries are 0 or 1,[12] with the property that any two rows share precisely $n/2$ entries. For $n = 2$, an example is

$$H := \begin{pmatrix} 1 & 1 \\ 1 & 0 \end{pmatrix}.$$

Here, the two rows share the first entry, but differ at the second position. For $n = 4$, an example is

$$\begin{pmatrix} 1 & 1 & 1 & 1 \\ 1 & 0 & 1 & 0 \\ 1 & 1 & 0 & 0 \\ 1 & 0 & 0 & 1 \end{pmatrix}.$$

---

[12] In the literature, more precisely, it is required that the entries be 1 or -1, but this leads to an equivalent problem.

Here, any two rows agree in precisely two positions, as required. For instance, the third and the fourth row share the first and the third entry. Construct a Hadamard matrix for $n = 2^k, k \in \mathbb{N}$. (By understanding how the example for $n = 4$ is constructed from the building block H for $n = 2$, you will probably rediscover a construction first found by Sylvester, a long time before Hadamard.) Try to find Hadamard matrices for other values of n.

It is conjectured that a Hadamard matrix exists if and only if $n = 2$ or $n = 4m$ for some $m \in \mathbb{N}$, but this is as yet unsettled.

2. Here is an easy exercise: Consider the graph whose vertices are the members of a population and whose edges represent matings between them (for simplicity, we do not consider the number of matings between the same pair, but only discuss the unweighted graph with an edge between two individuals that have mated at least once). What qualitative properties should this graph possess? How can you read off particular mating structures in the population, like polygamy or monogamy, polygyny (a male individual may have several mating partners, a female only one) or polyandry (the other way around)?

3. Another easy one: Argue that a trophic network, or put in simpler words, a food web, whose vertices are species in an ecosystem and an edge between two vertices expresses that one species feeds upon the other one, should be represented by a directed graph that does not contain directed cycles. Or should we admit exceptions? Estimate how long a directed path could maximally be (this is called the number of trophic levels—you may want to check the biological literature on this issue). Develop criteria in terms of the structure of this directed graph for assessing the importance of a particular species for an ecosystem.

4. List all non-isomorphic connected graphs with 5 vertices.

5. What is the smallest order for which there exists a graph without any nontrivial automorphism?

6. Determine the clustering coefficients and the k-cores of the following graphs and estimate their Polya-Cheeger constants,

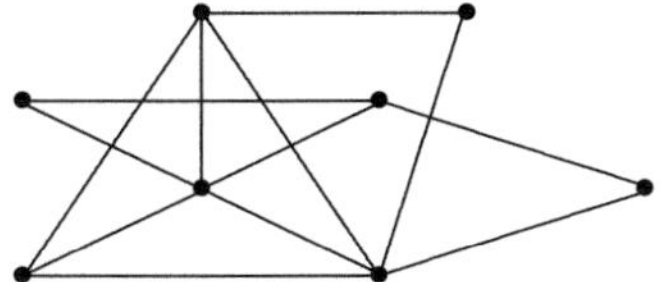 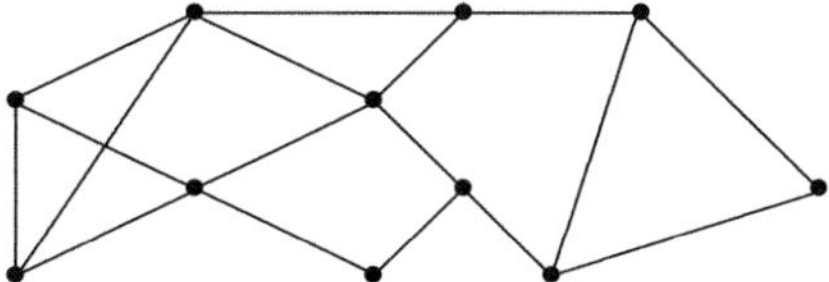

7. Let Γ be a k-regular graph with N vertices. Denote the eigenvalues of the adjacency matrix of Γ by $\mu_1 \leq \mu_2 \leq \cdots \leq \mu_N$. What is the relationship between the μ_j and the eigenvalues $0 = \lambda_0 \leq \lambda_1 \leq \cdots \leq \lambda_{N-1}$ of the normalized Laplacian of Γ?

8. (a) The N-cycle C_N is the graph of N vertices $\{i_1, \ldots, i_N\}$ where vertex i_k is connected with the vertices i_{k-1} and i_{k+1} mod N. Show that its eigenvalues are $1 - \cos \frac{2\pi j}{N}, j = 0, \ldots, N - 1$.

(b) The N-path P_N is obtained from the N-cycle C_N by cutting the link between the vertices i_1 and i_N. Show that the eigenvalues of P_N are $1 - \cos\frac{\pi j}{N-1}$, $j = 0, \ldots, N - 1$.

(c) Show that the eigenvalues of the N-cube Q_N (which has 2^N vertices) are $\frac{2j}{N}$, $j = 0, \ldots, N$, with multiplicities $\binom{N}{j}$.

(d) The m-petal graph has one central vertex i_0 and $2m$ peripheral vertices $i_1, \ldots, i_{2m}$ such that i_0 is connected with every other vertex and in addition, vertex i_{2j-1} is connected with vertex i_{2j} for $j = 1, \ldots, m$. Show that its eigenvalues are 0, $\frac{1}{2}$ with multiplicity $m - 1$, and $\frac{3}{2}$ with multiplicity $m + 1$.

9. Determine the spectra of the following graphs,

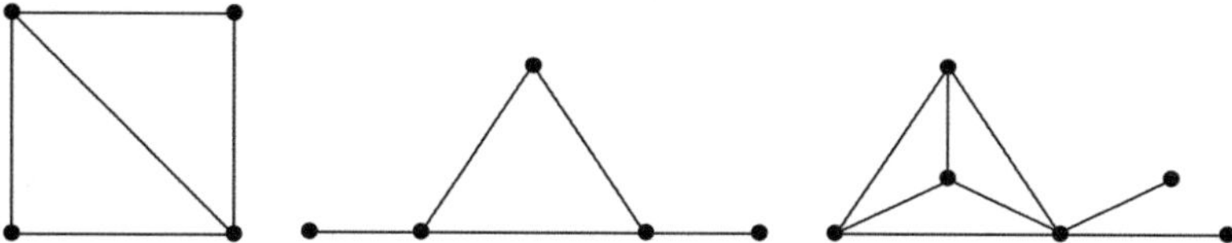

by using symmetries and/or node duplications.

10. Here is a more difficult exercise. Show that the following two graphs have the same spectrum, i.e., they are isospectral.

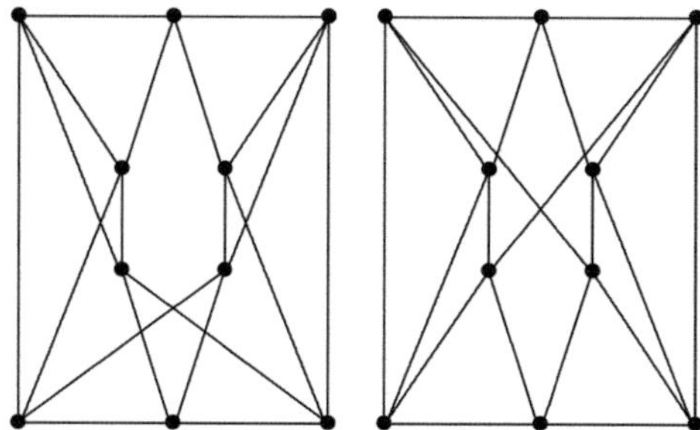

(Note: This example is taken from [118], but you need an additional step to solve this exercise, because in that reference, the spectrum of the adjacency matrix is studied instead of that of the Laplacian—see a preceding exercise for the relationship between the two.)

11. Take a graph Γ with an edge $i_1 \sim i_2$ and create a graph Γ' by duplicating that edge, i.e., add two vertices connected by an edge, $j_1 \sim j_2$, to Γ and connect j_1

to all neighbors of i_1, j_2 to all neighbors of i_2. What can you say about the effect on the spectrum? Determine the spectrum of the following example.

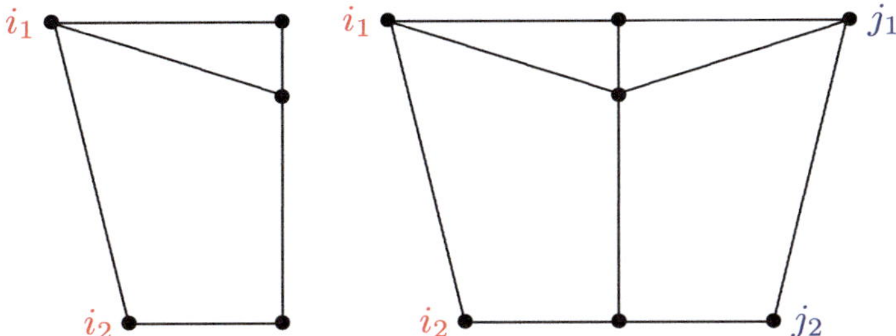

Also, observe that the m-petal graph described in one of the preceding exercises is obtained from the complete graph K_3 by successive edge duplications. Use this observation to explain its spectrum as computed in that preceding exercise.

12. What constraints does a pedigree graph in a bisexual population have to satisfy?

Chapter 3
Stochastic Processes

Abstract
Questions:

- How can the seemingly random firing pattern of a neuron encode any information about the inputs received?
- What will eventually happen to a population when the number of offspring of each individual randomly fluctuates?

We introduce the theory of stochastic processes. The coding and decoding of input information in systems of neurons is then modeled in terms of Poisson processes. Whereas in the last chapter we have treated descendence relations backward in time, to trace the ancestors, here we use branching processes to predict the future of populations.

3.1 Random Variables

A general reference for this chapter is [55].
Let (Ω, Σ, p) be a measure space, that is, a set Ω with a probability measure p and a Sigma algebra Σ of measurable sets. A measurable function

$$X : \Omega \to \mathbb{R} \tag{3.1.1}$$

then is called a random variable. The possible values of X are called events. Thus, what is random here is not the function X, but rather its argument $\omega \in \Omega$ that is considered to be drawn according to the measure p. The elements of Ω may represent the possible outcomes of some experiment or observation. When tossing a coin, for example, there are two possible outcomes, heads H and tails T, and these then are the elements of Ω. When the coin is tossed twice, the appropriate Ω contains 4 elements, HH, HT, TH, TT. The random variable may be the number of heads; in the last example, it may take the values 0, 1, and 2. In this situation, X takes only discrete

J. Jost, *Mathematical Methods in Biology and Neurobiology*, Universitext,
DOI: 10.1007/978-1-4471-6353-4_3, © Springer-Verlag London 2014

values, and whenever that is the case, we speak of a discrete random variable. When it takes its values in the non-negative integers $\mathbb{N} = \{0, 1, 2, \ldots\}$, we have a counting variable. This will be the important case for us. For a discrete random variable, we have the (probability) mass function

$$p(x) := p(X = x) := p(\{\omega \in \Omega : X(\omega) = x\}), \qquad (3.1.2)$$

with some abuse of notation (more precisely, we are using the symbol p both for the probability measure in Ω and for the probability measure in $\mathbb{R}$ induced by X—when there is a danger of confusion, we shall write p_X for the latter).

The distribution function of the random variable X is the function $f(x) := p(X \leq x)$. When $Y : \Omega \to \mathbb{R}$ is another random variable, we can consider the joint mass function

$$p(x, y) := p(X = x, Y = y) := p(\{\omega \in \Omega : X(\omega) = x, Y(\omega) = y\}) \qquad (3.1.3)$$

and the conditional one,

$$p(y|x) := p(Y = y|X = x) = \frac{p(X = x, Y = y)}{p(X = x)} \qquad (3.1.4)$$

whenever $p(X = x) > 0$, which satisfies

$$p(Y = y) = \sum_x p(Y = y|X = x)p(X = x) = \sum_x p(X = x, Y = y). \qquad (3.1.5)$$

The random variables X and Y are called independent if the probabilities for all events x of X and y of Y satisfy

$$p(X = x, Y = y) = p(X = x)p(Y = y). \qquad (3.1.6)$$

X and Y are called identically distributed if for every $z \in \mathbb{R}$

$$p(X = z) = p(Y = z). \qquad (3.1.7)$$

The abbreviation "i.i.d" meaning independent and identically distributed is frequently used.

In the above example, we may take X as the number of heads. When the coin is fair, that is, when HH, HT, TH, TT each occur with probability 1/4, we have $p(0) = 1/4, p(1) = 1/2, p(2) = 1/4$. More generally, when heads turn up with probability q, and if the results of the two tosses are independent of each other, we have $p(0) = (1 - q)^2, p(1) = 2q(1 - q), p(2) = q^2$. More generally, if we toss the coin n times, then $\binom{n}{k}$ points in the corresponding Ω, the set of the possible tossing sequences, yield k heads, and the mass function is

$$p(k) = \binom{n}{k} q^k (1-q)^{n-k}. \tag{3.1.8}$$

This is the binomial distribution $Bi(n, q)$. When we let $n \to \infty$, $q \to 0$ in such a manner that $nq \to \lambda \neq 0$, then, using the approximations $\binom{n}{k} \approx \frac{n^k}{k!}$ and $(1-q)^{n-k} \approx (1-q)^n = (1 - \frac{\lambda}{n})^n \approx e^{-\lambda}$, we obtain the limit

$$\binom{n}{k} q^k (1-q)^{n-k} \to \frac{\lambda^k}{k!} e^{-\lambda}. \tag{3.1.9}$$

This is the Poisson distribution $Q(\lambda)$.

The basic continuous distribution is the Gauss distribution, also called the normal distribution

$$\frac{1}{\sqrt{2\pi}\sigma} \exp\left(-\frac{(x-\mu)^2}{2\sigma^2}\right) \tag{3.1.10}$$

on $\mathbb{R}$ with mean μ and variance σ^2. Similarly, on $\mathbb{R}^n$, we have the multinomial Gauss distribution for a positive definite covariance matrix Σ

$$\frac{1}{\sqrt{(2\pi)^n |\Sigma|}} \exp\left(-\frac{1}{2}(x-\mu)^{\mathrm{T}}\Sigma^{-1}(x-\mu)\right), \tag{3.1.11}$$

for $x, \mu \in \mathbb{R}^n$.

Definition 3.1.1. The k-th moment of the discrete random variable X with mass function p is the expectation value of X^k,

$$E(X^k) = \sum_x x^k p(x). \tag{3.1.12}$$

whenever that sum converges absolutely.

In fact, in order to make this consistent, one verifies more generally that for $\phi : \mathbb{R} \to \mathbb{R}$

$$E(\phi(X)) = \sum_x \phi(x)p(x) \tag{3.1.13}$$

whenever the sum converges absolutely.

When the random variable X is not discrete, the above sums get replaced by an integral; for example

$$E(X^k) = \int_x x^k p(x)dx. \tag{3.1.14}$$

For the moment, however, we consider discrete random variables.

The first moment $E(X)$ is of course the mean, average, or expectation value of X, and the second moment then yields its variance $\text{var}(X) = E(X^2) - (E(X))^2 = E((X - E(X))^2)$.

We can also consider conditional expectations and obtain the following lemma:

Lemma 3.1.1. *Let X and Y be discrete random variables on Ω. The conditional expectation $E(Y|X) =: \phi(X)$ satisfies*

$$E(\phi(X)) = E(Y). \tag{3.1.15}$$

Proof. By (3.1.13) and (3.1.5)

$$E(\phi(X)) = \sum_x \phi(x)p_X(x) = \sum_x \sum_y y\, p_{Y|X}(y|x)p_X(x)$$
$$= \sum_y y\, p_Y(y) = E(Y).$$

$\square$

The random variables X and Y are called uncorrelated if

$$E(XY) = E(X)E(Y). \tag{3.1.16}$$

Independent random variables are uncorrelated, but not necessarily conversely. The next result is easy:

Lemma 3.1.2. *For random variables X, Y and $\alpha, \beta \in \mathbb{R}$, we have*

(a)
$$E(\alpha X + \beta Y) = \alpha E(X) + \beta E(Y). \tag{3.1.17}$$

(b)
$$var(\alpha X) = \alpha^2 var(X) \tag{3.1.18}$$

(c) when X and Y are uncorrelated

$$var(X + Y) = var(X) + var(Y). \tag{3.1.19}$$

$\square$

(a) simply expresses the linearity of the expectation value. Therefore, one may perform arbitrary *linear* operations with random variables without requiring that they be independent and have the corresponding operations on the expectation values.

We apply this lemma to the binomial distribution $Bi(n, q)$ from (3.1.8), interpreted as n independent tossings of a coin. Since the value of the random variable H, the number of heads, in a single toss is 0 or 1, we have $E(H^2) = E(H) = q$ and so the variance $E(H^2) - (E(H))^2$ is $q(1 - q)$. Thus, the random variable for the event H has expectation value q and variance $q(1 - q)$. We therefore obtain from the lemma

$$E(Bi(n, q)) = nq, \quad \text{var}(Bi(n, q)) = nq(1 - q). \tag{3.1.20}$$

For the Poisson distribution $Q(\lambda)$ (3.1.9) obtained above as a limit of binomial distributions, we obtain

$$E(Q(\lambda)) = \text{var}(Q(\lambda)) = \lambda. \tag{3.1.21}$$

There are various notions of convergence for a sequence $(X_n)_{n \in \mathbb{N}}$ of random variables towards a random variable X. X_n converges to X almost surely if $p(\omega \in \Omega : \lim_{n \to \infty} X_n(\omega) = X(\omega)) = 1$. It converges to X in kth mean ($k \geq 1$) if all $E(|X_n|^k) < \infty$ and $\lim E(|X_n - X|^k) = 0$. The cases of most interest are $k = 1$ and $k = 2$. For $k = 2$, one speaks of convergence in mean square. (X_n) converges to X in probability if for all $\varepsilon > 0$, $\lim_{n \to \infty} p(|X_n - X| > \varepsilon) = 0$ (as usual, the expression here is shorthand for $p(\{\omega \in \Omega : |X_n(\omega) - X(\omega)| > \varepsilon\}))$. It converges to X in distribution if $\lim_{n \to \infty} p(X_n \leq x) = p(X \leq x)$ for all $x \in \mathbb{R}$ for which the right hand side is continuous. Almost sure convergence and convergence in kth mean ($k \geq 1$) each imply convergence in probability, and the latter in turn implies convergence in distribution, and convergence in kth mean implies convergence in lth mean for $k > l \geq 1$. There exist no other general implications between these notions of convergence.

We now state some fundamental convergence theorems, referring to [55] for proofs. The first is the **strong law of large numbers**:

Theorem 3.1.1. *Let X_n, $n \in \mathbb{N}$ be i.i.d random variables with $E(|X_1|) < \infty$. With $\mu := E(X_1)$, we then have*

$$\frac{1}{n} \sum_{\nu=1}^{n} X_\nu \to \mu \text{ almost surely.} \tag{3.1.22}$$

If $E(X_1^2) < \infty$, the convergence takes also place in mean square.

("Strong" here refers to almost sure convergence, as opposed to the weak law of large numbers where convergence only takes place in probability.)

The next is the **central limit theorem**:

Theorem 3.1.2. *Let (X_n) be a sequence of i.i.d. random variables with finite mean μ and finite variance $\sigma^2 \neq 0$. Then, for $S_n := \sum_{\nu=1}^{n} X_\nu$, the distribution of*

$$\frac{S_n - n\mu}{\sqrt{n\sigma^2}} \text{ converges in distribution to the Gaussian distribution} \frac{1}{\sqrt{2\pi}} \exp(-\frac{x^2}{2}). \tag{3.1.23}$$

Let now

$$X : \Omega \to \mathbb{N} \tag{3.1.24}$$

be a discrete random variable that assumes only non-negative integer values.

Definition 3.1.2. The generating function of the random variable X is

$$G(s) := E(s^X) = \sum_{n=0}^{\infty} s^n p(n) \tag{3.1.25}$$

(defined for those values of $s \in \mathbb{R}$ for which the sum converges).

Of course, the sequence $p(n)$ can be recovered from the generating function by evaluating its kth derivative at 0:

$$G^{(k)}(0) = k!\, p(k) \tag{3.1.26}$$

As we shall now see, the derivatives of G at 1 also encode important properties of the sequence $p(n)$, namely its moments. Thus, when letting the argument s vary from 0 to 1, the generating function interpolates between the individual probabilities and collective properties of the distribution, the moments. Moreover, the generating function behaves in a very useful manner under composition of random processes and allows for a computation of moments of composed processes from the moments of the individual processes.

Lemma 3.1.3. *(a) Let G be the generating function of X. Then*

$$E(X) = G'(1) \text{ and more generally } E(X(X-1)\cdots(X-k+1)) = G^{(k)}(1) \tag{3.1.27}$$

whenever that k-th derivative of G at $s = 1$ exists. (Thus, the moments of X can be computed recursively from the generating function.)

(b) If $X_1, \ldots, X_N$ are independent random variables, their generating functions satisfy

$$G_{\sum_{\nu=1}^{N} X_\nu}(s) = G_{X_1}(s) \cdots G_{X_N}(s). \tag{3.1.28}$$

(c) If $X_1, X_2, \ldots$ are independent and identically distributed random variables which then have the same generating function, denoted by G_X, and if $N : \Omega \to \mathbb{N}$ is another random variable independent of the X_ν with generating function G_N, then the random variable $Y = X_1 + X_2 + \cdots + X_N$ (that is, the number of random variables occuring in the sum is now a random variable itself) has the generating function

$$G_Y(s) = G_N(G_X(s)). \tag{3.1.29}$$

Proof. (a) is obvious, and (b) follows from (3.1.16) applied to the independent random variables s^{X_ν}. For (c), from Lemma 3.1.1

$$G_Y(s) = E(s^Y) = E(E(s^Y|N)) = \sum_n E(s^Y|N=n)\,p_N(n)$$

$$= \sum_n E(s^{(X_1+\cdots+X_n)})\,p_N(n)$$

$$= \sum_n E(s^{X_1})\cdots E(s^{X_n})\,p_N(n) \text{ by (b)}$$

$$= \sum_n G_X(s)^n\,p_N(n) = G_N(G_X(s)).$$

$\square$

An alternative to the above polynomial generating function is the exponential one where we replace s in (3.1.25) by e^t to get

$$H(t) := E(e^{tX}) = \sum_{n=0}^{\infty} e^{nt}p(n). \tag{3.1.30}$$

Here, the moments of p can directly be computed from the derivatives of $H(t)$ at $t = 1$. All the formal results that we demonstrate about G also hold for H (see exercise 5.(a) below). In fact, for most purposes, the generating function H is more convenient than G. In Theorem 3.4.1 below, however, we shall need a particular property of G, and this is the main reason why are working here systematically with G in place of H. Also, it will be useful in our discussion of random graphs below. A variant of H is the discrete Fourier transform, the so-called characteristic function

$$E(e^{itX}) = \sum_{n=0}^{\infty} e^{int}p(n) = \sum_n \sum_{\nu=0}^{\infty} \frac{(it)^\nu}{\nu!} n^\nu p(n), \tag{3.1.31}$$

also called the moment generating function, that similarly encodes the properties of the distribution $p(n)$.

3.2 Random Processes

Definition 3.2.1. A random or stochastic process is a family $X = (X_t)$ of random variables indexed by some set $T \subset \mathbb{R}$.

Remark. More generally, one can allow for the X_t to take values in some measurable space other than $\mathbb{R}$.

As X_t is a random variable, for each t, we get an induced probability distribution for the values of X by

$$p_t(S) := p(X_t \in S) := p(\{\omega \in \Omega : X_t(\omega) \in S\}) \text{ for a measurable } S \subset \mathbb{R}, \tag{3.2.1}$$

analogously to (3.1.2) (where we had only considered the case of discrete values). The random variables X_{t_1} and X_{t_2} (say $t_1 < t_2$) are independent, see (3.1.6), if for $S_1, S_2 \subset \mathbb{R}$

$$p(X_{t_1} \in S_1, X_{t_2} \in S_2) = p(X_{t_1} \in S_1)\, p(X_{t_2} \in S_2). \tag{3.2.2}$$

The process is called stationary if its finite dimensional distributions are time invariant, i.e.

$$p(X_{t_1+\tau} \in S_1, ..., X_{t_n+\tau} \in S_n) = p(X_{t_1} \in S_1, ..., X_{t_n} \in S_n) \tag{3.2.3}$$

for all $t_1 < ... < t_n$, $S_1, ..., S_n \subset \mathbb{R}$ and $-\infty < \tau < \infty$.

We shall now consider the case $T = \mathbb{N}$.

Definition 3.2.2. The random process X is called a Markov chain if

$$p(X_{n+1} = x | X_1 = x_1, X_2 = x_2, \ldots, X_n = x_n) = p(X_{n+1} = x | X_n = x_n) \tag{3.2.4}$$

for all $n \geq 1$, $x, x_1, \ldots, x_n \in \mathbb{R}$.

Definition 3.2.3. The random process X is called a martingale if $E(|X_n|) < \infty$ for all n and

$$E(X_{n+1} | X_1, X_2, ..., X_n) = X_n. \tag{3.2.5}$$

We have the fundamental martingale convergence theorem

Theorem 3.2.1. *A martingale X with $E(X_n^2) < K < \infty$ for some K and all n converges to some random variable Ξ almost surely and in mean square.*

3.3 Poisson Processes and Neural Coding

In neurobiology, one studies spikes, that is, firings of neurons. Whereas there exist biophysical models for the generation of spikes on the basis of the electrochemical dynamics within neurons, see the Hodgkin-Huxley model described in Sect. 4.3.1 below, in more abstract models of information processing, spikes are conceptualized as discrete events occurring at time points, that is, events without temporal duration. This motivates

Definition 3.3.1. A stochastic process $(N_t)_{t \in T}$ is called a *point process* when T is an interval in $\mathbb{R}$ and the random variables N_t take only discrete values. It is called a *counting process* when $N_0(\omega) = 0$ and $N_t(\omega)$ counts the number of specified events in the interval $[0, t]$.

Of course, these concepts also apply to events other than spikes.
Obviously, one can relax the normalizations, e.g., take some other t_0 in place of 0, without much gain of generality.
Since the realization $\omega \in \Omega$ will not play a significant role in the sequel, we shall omit it in our notation and henceforth write

$$N(t) \text{ in place of } N_t(\omega).$$

One may also consider the probability distribution of the time interval between events. The combination of these two points of view, that is, counting the number of events having occured until time t vs. recording the temporal distance between subsequent events, will prove quite insightful.—If these time intervals are independently and identically distributed (a point to be returned to shortly), the process is called a *renewal process*. That is, the nth event occurs at time $t_1 + t_2 + \ldots + t_n$ where $t_1, \ldots, t_n$ are independent positive random variables that are identically distributed according to some probability density function $p(t)$. When this distribution depends only on the difference between time points, but not otherwise on time, the process is called homogeneous, else inhomogeneous. We return to the counting process.

Definition 3.3.2. A counting process $N(t)$ is called locally continuous in probability if

$$\lim_{\epsilon \searrow 0} p(N(t + \epsilon) - N(t) \geq 1) = 0. \tag{3.3.1}$$

Definition 3.3.3. A counting process $N(t)$ is said to have independent increments if the numbers of events in disjoint time intervals are independent.

Definition 3.3.4. A counting process $N(t)$ is said to have stationary increments if the number of events in a time interval depends only on the length of that time interval.

Definition 3.3.5. A counting process $N(t)$ that is locally continuous in probability and has independent and stationary increments is called a (homogeneous) Poisson process.

Theorem 3.3.1. *For a Poisson process $N(t)$*

$$p(N(s + t) - N(s) = n) = e^{-rt} \frac{(rt)^n}{n!} \tag{3.3.2}$$

for $n = 0, 1, \ldots$, for some constant $r \geq 0$.

Thus, the number of events produced by a Poisson process is distributed according to the Poisson distribution $Q(rt)$, see (3.1.9). The parameter rt of that distribution thus is proportional to the length t of the time interval, and the factor r is called the rate of the process.

Proof. We put

$$\rho(t) := p(N(t) - N(0) \geq 1) = p(N(s + t) - N(s) \geq 1) \tag{3.3.3}$$

where the last equality holds because N has stationary increments. We claim that

$$\rho(t) = 1 - e^{-rt} \tag{3.3.4}$$

for some constant $r \geq 0$. To see this, we start by observing that, by the assumption of independent increments, the probability $\pi_0(t + s)$ that no event occurs between 0 and $t + s$ equals the product of the probabilities that no event occurs in $[0, t]$ and that no event occurs in $[t, t + s]$. By the assumption of stationary increments, the latter equals the probability that no event occurs in $[0, s]$, and hence

$$\pi_0(t + s) = \pi_0(t)\pi_0(s) \tag{3.3.5}$$

whence $\pi_0(t) = \pi_0(0) \exp(-rt)$ for some constant r. $r \geq 0$ since this probability is decreasing in t. Since $\pi_0(0) = 1$ as $N(0) = 0$, we obtain

$$\pi_0(t) = e^{-rt}. \tag{3.3.6}$$

Since this is the probability that no event occurs until t, the probability $\rho(t)$ that the first event occurs before t is $1 - e^{-rt}$ which is (3.3.4).
Then

$$\rho(t) = rt + o(t) \ \text{ for } \ t \to 0. \tag{3.3.7}$$

Similarly, the probability $\pi_1(t) := p(N(t) = 1)$ that precisely one event occurs between 0 and t satisfies

$$\pi_1(t + s) = \pi_0(t)\pi_1(s) + \pi_1(t)\pi_0(s) = e^{-rt}\pi_1(s) + e^{-rs}\pi_1(t). \tag{3.3.8}$$

Using (3.3.7), we obtain

$$\pi_1(t) = rte^{-rt}. \tag{3.3.9}$$

Iteratively, we obtain

$$\pi_n(t) := p(N(t) = n) = \frac{(rt)^n}{n!}e^{-rt}. \tag{3.3.10}$$

Recalling the assumption of stationary increments, this shows (3.3.2).   □

As a consistency check, the reader should verify that a counting process $N(t)$ given by (3.3.2) with independent increments conversely has stationary increments and is locally continuous in probability.

We also observe that

$$p(N(t) < \infty) = \sum_{n \geq 0} p(N(t) = n) = e^{-rt} \sum_{n \geq 0} \frac{(rt)^n}{n!} = 1, \tag{3.3.11}$$

and therefore, almost surely, only finitely many events take place in the finite interval $[0, t]$. Alternatively, this can also be directly deduced from the assumptions. If there were infinitely many events in $[0, t]$, then there would also be infinitely many events in either $[0, t/2]$ or $[t/2, t]$. By the assumption of stationary increments, then there would be infinitely many events in *both* these intervals. Repeating this argument, there would be infinitely many events in every subinterval of $[0, t]$. Consequently, every point would be an accumulation point of events, contradicting the assumption of local continuity in probability.

Theorem 3.3.2. *For a Poisson process with rate r, the time of the nth event is distributed according to*

$$p_n(t) = \frac{r(rt)^{n-1}}{(n-1)!} e^{-rt}. \tag{3.3.12}$$

Proof.   We have

$$p(N(t) = n) = \int_0^t p_n(s)ds - \int_0^t p_{n+1}(s)ds, \tag{3.3.13}$$

the probability that the nth event occurs in $[0, t]$ minus the probability that the $(n + 1)$st event occurs in that interval. Hence

$$\frac{d}{dt} p(N(t) = n) = p_n(t) - p_{n+1}(t). \tag{3.3.14}$$

Since $p_0(t) = 0$ for $t > 0$, (3.3.12) follows iteratively from (3.3.10).   □

The argument of the proof can also be reversed:

$$\int_0^t p_1(s)ds = \rho(t), \tag{3.3.15}$$

hence the probability of observing the first event at t is given by the derivative of the probability for the first event occurring before t. Thus, $p_1(t) = ce^{-rt}$ for some constant c, and (3.3.4) gives us $c = r$. Thus, we obtain

$$p_1(t) = re^{-rt}, \tag{3.3.16}$$

and as in the proof of Theorem 3.3.1, we can iterate that argument.
We also have the consistency relation

$$\sum_{\nu=0}^{n-1} \frac{(rt)^\nu}{\nu!} e^{-rt} = \int_t^\infty \frac{r(r\tau)^{n-1}}{(n-1)!} e^{-r\tau} d\tau \tag{3.3.17}$$

(which follows from integration by parts and induction in n). The lhs is the probability that at most $n-1$ events have taken place up to time t (Theorem 3.3.1), and the rhs that the nth event occurs after that time (Theorem 3.3.2).
Assume that $N(t) = n$, that is, precisely n events occur in $[0, t]$. Since by the assumptions of stationary and independent increments, every n-tuple $(t_1, \ldots, t_n)$ of points in $[0, t]$ has the same probability density of receiving those n events, that probability density is given by

$$p(t_1, ..., t_n) = \frac{1}{n!} r^n e^{-rt} \tag{3.3.18}$$

because this depends on the length t of the interval and yields (3.3.2) by integration w.r.t. t_1, ..., t_n from 0 to t. The combinatorial factor $\frac{1}{n!}$ arises, because the n events under consideration are indistinguishable. This also leads to the correct normalization, because when we integrate w.r.t. t_1, ..., t_n and then sum over all n, we obtain 1, by the relation

$$\sum_n \frac{1}{n!} (rT)^n = e^{rT}. \tag{3.3.19}$$

Theorem 3.3.3. *Let $N_1(t), N_2(t)$ be independent Poisson processes with rates r_1, r_2. Then the counting process $N(t) = N_1(t) + N_2(t)$ is a Poisson process with rate $r = r_1 + r_2$.*

The *proof* is an obvious verification of the defining properties of a Poisson process. $\qquad\qquad\square$
Likewise, we can multiply the rate of a Poisson process with a constant positive factor to obtain another Poisson process. In fact, when the factor is not constant, but is a function of time, we still obtain a process that is not significantly different as we shall now explore.

Definition 3.3.6. Let $r(t)$ be a continuous positive function on $\mathbb{R}_+$, and put $r_t := \int_0^t r(\tau)d\tau$. A counting process $N(t)$ with independent increments that satisfies

$$p_t(n) := p(N(s+t) - N(s) = n) = e^{-r_t} \frac{(r_t)^n}{n!} \tag{3.3.20}$$

for $n = 0, 1, \ldots$ is called an (inhomogeneous) Poisson process with rate function $r(t)$.

Actually, this is not such a vast generalization (and we indicated that already by the typographical similarity between rt and r_t). Since the parametrization of time is arbitrary, we have

Theorem 3.3.4. *Any inhomogeneous Poisson process can be transformed into a homogeneous one via a time reparametrization.*

Proof. Since $r(t)$ is assumed positive, r_t is a strictly monotonically increasing, continuous function of t with $r_0 = 0$ and $\lim_{t \to \infty} r_t = \infty$. It therefore has an inverse function $\rho(t)$ with the same properties. We put

$$N'(t) := N(\rho(t)). \tag{3.3.21}$$

This counting process then is a homogeneous Poisson process with rate 1 since $r_{\rho(t)} = t$. $\qquad\square$

The assumption that $r(t)$ be strictly positive could be relaxed to nonpositivity. In that case, the time transformation would simply jump over those periods where $r(t)$ vanishes.

An equivalent characterisation of an inhomogeneous Poisson process with rate function $r(t)$ is that it be a counting process with independent increments, satisfying

$$p(N(t+\epsilon) - N(t) = 1) = r(\epsilon) + o(\epsilon) \text{ and } p(N(t+\epsilon) - N(t) \geq 2) = o(\epsilon) \text{ for } \epsilon \to 0. \tag{3.3.22}$$

The first relation is a quantitative version of the local continuity in probability. The second relation automatically holds in the homogeneous case, but one needs to assume this in the inhomogeneous case.

For an inhomogeneous Poisson process, the probability density for observing events at the times $t_1, \ldots, t_n$ is

$$p(t_1, \ldots, t_n) = \frac{1}{n!} \exp\left(-\int_0^t r(\tau)d\tau\right) \prod_{i=1}^{n} r(t_i). \tag{3.3.23}$$

Using the time reparametrization of Theorem 3.3.4, this is deduced from the formula (3.3.18) for the homogeneous case. Upon integration, it yields (3.3.20), analogously to the homogeneous case.

From (3.3.20), the expected number of events in the interval $[0, t]$ is

$$\sum_n n p_t(n) = \sum_n \frac{(r_t)^n}{(n-1)!} e^{-r_t} = r_t = \int_0^t r(\tau)d\tau \tag{3.3.24}$$

(using (3.3.19) for the second equality).

Inhomogeneous Poisson processes are important models in theoretical neurobiology (see e.g. [33]) because the rate $r(t)$ that represents the spiking or firing rate of a neuron can then depend on the stimulus S, that is, on the input received by the neuron. Thus,

$$r(t) = r(t; S), \tag{3.3.25}$$

and this represents the coding scheme of the neuron under consideration. In other words, when receiving the stimulus S, the neuronal firing follows a Poisson process with rate $r(t; S)$. For simplicity, we may assume that the stimulus is received at time $t = 0$. A basic model assumes that the neuron has a preferred or optimal stimulus S_0, and that the stimulus is translated via a Gaussian tuning function into the firing rate

$$r(t; S) = c \exp\left(-\frac{d^2(S, S_0)}{2\sigma^2}\right). \tag{3.3.26}$$

Here, $d(.,.)$ is some metric in the input space, for example a Euclidean one, that is, $d(S, S_0) = \|S - S_0\|$; c and σ^2 (variance) are parameters. Obviously, other coding schemes are possible.

From (3.3.23), we see that the probability distribution for observing spikes precisely at the times $t_1, ..., t_n$ in the interval $[0, t]$, given the input S is

$$p(t_1, ..., t_n | S) = \frac{1}{n!} \exp\left(-\int_0^t r(\tau; S)d\tau\right) \prod_{i=1}^{n} r(t_i; S). \tag{3.3.27}$$

Bayes' formula then yields the fundamental relationship for decoding the spike train produced by the neuron, that is, an estimate for the distribution of signals contingent upon the recorded spike train $t_1, \ldots, t_n$,

$$p(S|t_1, ..., t_n) = p(t_1, ..., t_n|S) \frac{p(S)}{p(t_1, ..., t_n)}. \tag{3.3.28}$$

Here, $p(S)$ is a prior estimate for the distribution of stimuli (that may have been obtained as the result of some learning process). $p(t_1, ..., t_n)$ simply represents some normalization factor.

Of course, (3.3.27) can also be used for other estimation schemes. For example, maximum likelihood selects a stimulus $\bar{S}$ that has caused an observed spike sequence with the highest probability, that is, $\bar{S} = \text{argmax } p(t_1, ..., t_n|S)$ in (3.3.27).

Clearly, a biological neuron does not operate according to a Poisson process. A spike is not an instantaneous event, but the generation of an action potential has a positive, although rather short, duration. Also, two spikes cannot be fired in too rapid succession because after a spike is fired, a neuron goes through a refractory period before it can generate the next spike. Thus, if one wants to include these aspects, one needs a biophysical in place of a purely phenomenological model. Such models exist, and we shall introduce and investigate below the basic one, the Hodgkin-Huxley model. Notwithstanding its lack of biophysical realism, however, Poisson type models are very important in the neurosciences because, on one hand, they relate well to the experimental practice of recording spikes, and on the other hand, they can be the basis for models of information transmission in neural systems.

3.4 Branching Processes

References for this section are [80, 56].

We start with the simplest branching process, the Galton-Watson process. Here, each individual lives in a fixed generation n and independently of all other individuals produces a random number of offspring that become members of generation $n + 1$. This random variable, the number of offspring, is the same for all individuals in all generations. Thus, the numbers of offspring for the individuals are independent and identically distributed random variables. We denote their common generating function by $G(s)$. We also assume that there is a positive probability for having more than one offspring. If the probability of having m offspring is $p(m)$, this means that $p(0) + p(1) < 1$.

Let the random variable Z_n denote the size of generation n. One usually assumes that the process starts with a single individual in generation 0, that is, $Z_0 = 1$. Let $G_n(s) = E(s^{Z_n})$ be the generating function of Z_n.

Lemma 3.4.1. *G_n is the n-th fold iterate of G,*

$$G_n(s) = G \circ \cdots \circ G(s), \qquad (3.4.1)$$

and thus also for $m, n \in \mathbb{N}$

$$G_{m+n}(s) = G_m(G_n(s)). \qquad (3.4.2)$$

Proof. We shall show (3.4.2) which easily implies (3.4.1) by iteration. Let the random variable Y_i denote the number of members of the $(m + n)$th generation that derive from member i of the mth one. We then have

$$Z_{m+n} = Y_1 + \cdots + Y_{Z_m}. \qquad (3.4.3)$$

By our assumptions, the Y_i are independent and identically distributed, in fact identical to Z_n, the number of offspring deriving from an individual n generations ago. Lemma 3.1.3 c) then yields the claim. $\square$

Corollary 3.4.1. *Let $\mu := E(Z_1)$ and $\sigma^2 := var(Z_1)$. Then*

$$E(Z_n) = \mu^n \text{ and } var(Z_n) = \begin{cases} n\sigma^2 & \text{if } \mu = 1 \\ \frac{\sigma^2(\mu^n - 1)\mu^{n-1}}{\mu - 1} & \text{if } \mu \neq 1. \end{cases} \tag{3.4.4}$$

Proof. Differentiating $G_n(s) = G(G_{n-1}(s))$ at $s = 1$ yields

$$E(Z_n) = \mu E(Z_{n-1}) \tag{3.4.5}$$

from which the first equation follows by iteration. Differentiating twice gives $G_n''(1) = G''(1)G_{n-1}'(1)^2 + G'(1)G_{n-1}''(1)$ which yields the second equation. $\square$

In view of this result, we call the process subcritical, critical, supercritical when $\mu < 1, = 1, > 1$, resp. In the sub- (super-)critical, we thus expect the population to shrink (grow) while in the critical it is expected to stay the same. This might lead one to expect that the population will continue forever, but that is not true as we shall now find out from asking whether the population will eventually become extinct, that is

$$Z_n = 0 \text{ for some } n \in \mathbb{N}, \text{ and then of course also for all } m \geq n. \tag{3.4.6}$$

Some observations are obvious:

- If $p(0) = 0$, that is, if every individual always has at least one offspring, then the population cannot become extinct. Therefore, we shall assume now

$$p(0) > 0. \tag{3.4.7}$$

- When $p(\nu) = 0$ for $\nu \geq 2$, then $p(0) + p(1) = 1$. We have excluded $p(0) = 0$ and therefore cannot have the trivial case $p(1) = 1$, that is, that every individual has precisely one offspring so that the population size will always remain constant. Consequently, the population should also become extinct because every individual then has either no or one offspring, and the former with a positive probability. Therefore, the population will certainly decrease. Therefore, we shall assume now

$$p(0) + p(1) < 1. \tag{3.4.8}$$

Theorem 3.4.1. *The extinction probability q_{ext} of the process, that is, the probability that for some $n \in \mathbb{N}$ we have $Z_n = 0$, equals the smallest root of the equation $G(s) = s$. For $\mu \leq 1$, we have $q_{ext} = 1$, that is, the population becomes extinct almost surely, while for $\mu > 1$, we have $q_{ext} < 1$, that is, the population has a positive probability of surviving forever.*

Proof. We observe that $G(s)$ as a power series with non-negative coefficients $p(\nu)$ and $p(0) + p(1) < 1$ by our initial assumptions is increasing and strictly convex for $s \in [0, 1]$ and satisfies $G(0) = p(0), G(1) = 1$. When $\mu = G'(1) \leq 1$, then

$G(s) > s$ for $s \in [0, 1)$, while for $\mu > 1$, $G(s) = s$ has a unique root in $[0, 1)$. These properties implies that $G(s) = s$ has a smallest root which we denote by q, and $q = 1$ for $\mu \leq 1$, but $q < 1$ for $\mu > 1$. Moreover, for $s \in [0, q)$, in particular for $s = 0$, the iterates $G_n(s) = G \circ \cdots \circ G(s)$ increase monotonically towards q while for $s \in (q, 1)$ it decreases monotonically towards q for $n \to \infty$. (In terms of dynamical systems, this simply expresses the stability of the fixed point of G at q under dynamical iteration of G, which always holds when the graph of a function G intersects the diagonal from above at a fixed point.) We recall from Lemma 3.4.1 that G_n is the generating function for Z_n. Thus

$$q = \lim_{n \to \infty} G_n(0) = \lim_{n \to \infty} p(Z_n = 0) = \lim_{n \to \infty} p(Z_\nu = 0 \text{ for some } \nu \leq n)$$

$$= p(Z_\nu = 0 \text{ for some } \nu \in \mathbb{N}) = p(\lim Z_n = 0)$$

is the extinction probability. $\square$

Thus, we see that due to the fluctuations in the number of offspring, a finite population may become extinct in finite time. It will do so almost surely when the expected number of offspring is at most 1—even when it is 1—, and it will also go extinct with a positive probability when that number is larger than 1. One may consider this as a finite size effect, in the sense that when we go to the limit of large populations (under appropriate technical conditions), the random fluctuations will average out and the population will expand or shrink deterministically at the rate μ.

A Galton-Watson branching process (Z_n) is a Markov process by (3.4.5). The normalized process $W_n := \frac{Z_n}{E(Z_n)}$ then is a martingale by (3.4.4),

$$E(W_{n+1}|W_1, \ldots, W_n) = W_n. \tag{3.4.9}$$

We now look at the situation where the expectation values $\mu(n)$ for the number of offspring of an individual in generation n vary, i.e., μ is a random variable itself (defined on $\mathbb{N}$). This is called a branching process in a random environment. The population then grows from time 0 to time n by the sequence $\mu(0), \mu(1), \ldots, \mu(n-1)$ which is equivalent to growth by the geometric mean $(\mu(0) \cdots \mu(n-1))^{1/n}$ in one step. In order to apply the law of large numbers, we need to convert this product into a sum,

$$(\mu(0) \cdots \mu(n-1))^{1/n} = \exp(\frac{1}{n}(\log \mu(0) + \cdots + \log \mu(n-1)) \tag{3.4.10}$$

and conclude by the law of large numbers

$$\lim_{n \to \infty} \frac{1}{n}(\log \mu(0) + \cdots + \log \mu(n-1)) = E(\log \mu) \tag{3.4.11}$$

with probability 1. Thus, the asymptotic population growth rate is

$$\lim_{n \to \infty} (\mu(0) \cdots \mu(n-1))^{1/n} = \exp(E(\log \mu)) \tag{3.4.12}$$

which may be smaller than $E(\mu)$. In particular, even when the latter may be >1, the process may still be subcritical, that is become extinct with positive probability, because of the fluctuations in the environment. So, once more we see that when a finite population is subjected to random effects its extinction probability may increase even when the expected growth rate stays the same.

The Galton-Watson process is the simplest branching process, and many generalizations are possible. One of them is to allow for individuals of different types $j = 1, \dots, m$. For each type, the distribution of the types of its offspring may be different. We then consider the matrix $M = (m_{ij})$ where m_{ij} is the expected number of offspring of type j of an individual of type i.[1] All entries of M are non-negative. The expectation value of the number $Z_{j,n}$ of individuals of type j in generation n is then

$$E(Z_{j,n}) = \sum_{i=1}^{m} E(Z_{i,n-1}) m_{ij} \tag{3.4.13}$$

by linearity. In vector notation, with $E(Z_n) = (E(Z_{1,n}), \dots, E(Z_{m,n}))^T$ (^{T} denoting transpose),

$$E(Z_n)^T = E(Z_{n-1})^T M = E(Z_0)^T M^n. \tag{3.4.14}$$

Of course, when as before, we specify the initial population Z_0, we can drop the last expectation E to get $E(Z_n)^T = Z_0^T M^n$.
We shall now apply the theory of Perron-Frobenius to the non-negative matrix M. That theory is summarized in

Lemma 3.4.2. *Let M be an $m \times m$ matrix with non-negative entries which is irreducible in the sense that all the entries of M^ν are even positive for some $\nu \in \mathbb{N}$. Then M has a simple eigenvalue ρ that is real and positive and larger than the absolute value of any other eigenvalue. It possesses a left eigenvector $u = (u^1, \dots, u^m)^T$ (that is, $u^T M = \rho u^T$, or in components, $\sum_j u^j m_{ji} = \rho u^i$) and a right eigenvector $v = (v^1, \dots, v^m)^t$ (i.e., $Mv = \rho v$) that both have positive entries. We can normalize them by*

$$\sum_{j=1}^{m} u^j = 1, \quad \sum_{j=1}^{m} u^j v^j = 1. \tag{3.4.15}$$

With $M_0 := (v^i u^j)_{i,j=1,\dots,m}$, we then have

---

[1] Later on, we shall consider the probability $p^i(n_1, \dots, n_m)$ that an individual of type i produces n_k offspring of type k. Then $m_{ij} = \sum_{n_1,\dots,n_m=1}^{\infty} n_j p^i(n_1, \dots, n_m)$.

$$M^n = \rho^n M_0 + \tilde{M}^n, \tag{3.4.16}$$

with $M_0\tilde{M} = \tilde{M}M_0 = 0$ and $|\tilde{M}^n| \leq const\ \tilde{\rho}^n$ for some $\tilde{\rho} < \rho$. In particular, for any non-trivial w with non-negative entries, the iterates $w^T M^n$ grow like ρ^n in norm.

Returning to (3.4.14), we see that when $E(Z_0) = cu$, i.e., is a multiple of the left eigenvector u for the maximal eigenvalue ρ, we obtain

$$E(Z_n)^T = cu^T M^n = cu^T \rho^n. \tag{3.4.17}$$

We conclude that the process is subcritical, critical or supercritical depending on whether $\rho < 1, = 1, > 1$. In fact, there seems to be a small caveat here, namely that for (3.4.17), we had assumed that the initial composition Z_0 of the population is a multiple of the positive eigenvector u. In order to be able to drop that assumption, we now assume that the matrix M is irreducible as in the Perron-Frobenius theorem. Such a process is called indecomposable. Then the value of ρ determines the asymptotic behavior of the population size for any initial configuration of the population.
We can also set up the generating function formalism as before, with the sole difference that all expressions now become vectors in place of scalars. The generating vector is $G = (G^1, \ldots, G^m)$ with

$$G^j(s_1, \ldots, s_m) = \sum_{n_1, \ldots, n_m} s_1^{n_1} \cdots s_m^{n_m} p^j(n_1, \ldots, n_m) \tag{3.4.18}$$

where $p^j(n_1, \ldots, n_m)$ is the probability that an individual of type j produces n_i offspring of type i, for $i = 1, \ldots, m$. We also have a vector $q = (q_1, \ldots, q_m)$ where q_j is the extinction probability for a population starting with a single individual of type j. Then, as in Theorem 3.4.1, the vector q is determined as the componentwise smallest root of the vector fixed point equation

$$q = G(q). \tag{3.4.19}$$

With some simple tricks, many different processes can be captured by multi-type Galton-Watson processes:

1. Given a single-type Galton-Watson process, we want to know the total number of individuals up to time n. We then simply define a second type of individual in the original process, the dead type. Type 1 corresponds to the original one, and it produces offspring of that type 1 according to the original rule, plus 1 individual of type 2, that is, it dies, as already assumed in the original process. An individual of type 2 produces one offspring of type 2, that is, it stays dead. By this token, the individuals from previous generations remain in all future generations, simply as corpses, i.e. as type 2 individuals. The transition matrix then is

$$\begin{pmatrix} \mu & 1 \\ 0 & 1 \end{pmatrix},$$

with $\mu = E(Z_1)$ as above.

2. We consider a population consisting of two sexes. Females produce offspring of either type, with equal probabilities, while males do not reproduce. Our transition matrix then is of the form

$$\begin{pmatrix} \mu/2 & \mu/2 \\ 0 & 0 \end{pmatrix}.$$

While this process again is decomposable, here the two types still grow or shrink at the same rate. In fact, however, this is only an incomplete model of populations with sexual reproduction because they can become extinct not only for the reason that the population size goes to 0, but also when one of the two sexes disappears. That aspect needs to be modeled separately through the choice of a mating function, that is, by a rule how the number of offspring depends on the numbers of the two sexes in the population.

3. One can also include dependencies between siblings. That means that the expected numbers and types of offspring of an individual depend not only on its own type, but also on that of its siblings. While this violates the independence hypothesis in Cor.3.4.1, it turns out that dependencies in the same generation do not affect the expected population size. Again, this can be seen through a simple trick, namely by formally considering the sibship (brood, litter) as the individuals in the process. Different such sibships then produce different sibships in the next generation.
We consider the case of altruistic siblings, for example where the eldest one may forego its own offspring for helping his younger siblings to raise additional offspring. Of course, the Galton-Watson assumptions that reproduction takes place at discrete time steps, and that each individual can reproduce only at age 1 make the distinction between older and younger siblings impossible if taken literally, but for the sake of the argument we assume that some litters contain an altruistic member—which we then simply label as the "eldest" —whereas others don't. In other words, we have two types of litters, one with an altruistic member, and the other one without. As before, m_{ij} is the expected number of progeny of type j produced by a litter of type i. By the Perron-Frobenius theorem, if M is irreducible, it has an eigenvalue ρ of largest absolute value that is positive and simple. ρ describes the growth rate, the corresponding normalized left eigenvector u yields the relative asymptotic contributions of the types to future generations while the right eigenvector v describes the asymptotic distribution of the process in case $\rho > 1$ in the following sense: We assume that the process is regular, i.e. that the probability for a litter producing more than one progeny is positive, and that the second moments of progeny distribution are finite. Then by the martingale convergence theorem 3.2.1, see (3.4.9), Z_n/ρ^n converges to a vector w which (with probability 1) is a positive (except in obvious trivial cases) multiple of v. In that case, the process will not become extinct with positive probability while in case $\rho \leq 1$, it goes extinct with probability 1 as is typical for branching processes. In

fact, in our simple situation with two types, that eigenvalue is given by

$$\rho = \frac{1}{2}(m_{11} + m_{22}) + \sqrt{\frac{1}{4}(m_{11} - m_{22})^2 + m_{12}m_{21}}. \tag{3.4.20}$$

Let us assume that type 1 is the altruistic, and type 2 the non-altruistic one, and that $m_{21} = 0$, i.e., that a non-altruistic cannot produce an altruistic one (M then is no longer irreducible but the needed results from the Perron-Frobenius theorem still hold here). Then, if $m_{11} > m_{22}$, i.e. the altruistic litters reproduce more successfully than the other ones,

$$\rho = m_{11}, \tag{3.4.21}$$

and the corresponding eigenvector is proportional to

$$\begin{pmatrix} m_{11} - m_{22} \\ m_{12} \end{pmatrix}. \tag{3.4.22}$$

Thus, the non-altruists only survive as a nontrivial fraction of the total population[2] if $m_{12} > 0$, that is if they are also produced by the altruists. Of course, this is rather obvious.

If altruism is caused by a single gene, then one can use the methods of mathematical population genetics to compute the transmission probability of the responsible allele in a sexually reproducing population. The point of our example is that the altruistic allele has to be present in the sibling labeled "eldest" for being effective, but that it can only be transmitted to the next generation when also present in his siblings for whose own behavior the allele is irrelevant. In this manner, the coefficient m_{12} can be determined.

3.5 Random Graphs

Equipped with tools from stochastic analysis, we now return to graph theory and discuss stochastic constructions of graphs. A good reference that we shall partly follow is [94].

The idea of Erdös and Rényi [39] that started the whole field was to not specify a graph explicitly, but rather only its generic type by selecting edges between nodes randomly, depending on a single parameter, the edge probability p. In a random graph, for any pair of nodes, there is thus an edge between them with probability p. If the network has N nodes, then each node has $N - 1$ possible recipients for an edge. Thus, the average degree of a node is

---

[2] Here, we are considering the population of litters, and not of individuals. For the latter, one would need to multiply these coefficients by the litter sizes.

$$z := (N - 1)p. \tag{3.5.1}$$

The case of interest is where $N \gg 1$, $p \ll 1$, and z is of intermediate size, that is neither 0 nor very large. Moreover, the probability that a given node has degree k in an Erdös-Rényi graph is

$$p_k = \binom{N-1}{k} p^k (1-p)^{N-1-k} \tag{3.5.2}$$

because the degree happens to be k when precisely k out of the $N-1$ possible edges from the given node are chosen, and each of them is chosen with probability p and not chosen with probability $1 - p$. Thus, the degree distribution is binomial, and for $N \gg kz$, this is approximated by the Poisson distribution

$$p_k = \frac{z^k e^{-z}}{k!}. \tag{3.5.3}$$

(and so $z = \langle k \rangle = \sum_k k p_k$ (cf. (3.1.9), (3.1.21) on the Poisson distribution and (3.1.27) on the generating function).)

For an Erdös-Rényi graph, one can also compute the distribution of the number of second neighbors of a given node, that is, the number of neighbors of its neighbors, discarding of course the original node itself as well as all its direct neighbors that also happen to be connected with another neighbor. However, since there is no tendency to clustering in the construction, the probability that a second neighbor is also a first neighbor behaves like $1/N$ and so becomes negligible for large N. Now, however, the degree distribution of first order neighbors of some node is different from the degree distribution of all the nodes in the random graph, because the probability that an edge leads to a particular node is proportional to that node's degree so that a node of degree k has a k-fold increased chance of receiving an edge. Therefore, the probability distribution of our first neighbors is proportional to $k p_k$, that is, given by $\frac{k p_k}{\sum_l l p_l}$, instead of p_k, the one for all the nodes in the graph. Since such a first neighbor of degree has $k - 1$ edges leading away from the original node, the distribution for having k second neighbors via one particular neighbor is then given, after shifting the index by 1, by

$$q_k = \frac{(k+1) p_{k+1}}{\sum_l l p_l}. \tag{3.5.4}$$

Thus, to obtain the number of second neighbors, we need to sum over the first neighbors, since, as argued, we can neglect clustering in this model. Thus, the mean number of second neighbors is obtained by multiplying the expected number of second neighbors via a particular first neighbor, that is, $\sum k q_k$, by the expected number of first neighbors, $z = \sum k p_k$. So, we obtain for that number

$$\sum_l l p_l \sum_k k q_k = \sum_{k=0}^{\infty} k(k+1)p_{k+1} = \sum_{k=0}^{\infty}(k-1)k p_k = \langle k^2 \rangle - \langle k \rangle. \qquad (3.5.5)$$

We recall from Sect. 3.1 that such probability distributions can be encoded in probability generating functions (see (3.1.25)). If we have a probability distribution p_k as above on the non-negative integers, we have the generating function (cf. (3.1.25))

$$G_p(x) := \sum_{k=0}^{\infty} p_k x^k. \qquad (3.5.6)$$

Likewise, the above distribution for the number of second neighbors then is encoded by

$$G_q(x) = \sum_{k=0}^{\infty} q_k x^k = \frac{\sum_k (k+1)p_{k+1}x^k}{\sum_l l p_l} = \frac{G_p'(x)}{z}. \qquad (3.5.7)$$

When we insert the Poisson distribution (3.5.3), we obtain

$$G_p(x) = e^{-z} \sum_{k=0}^{\infty} \frac{z^k}{k!} x^k = e^{z(x-1)} \qquad (3.5.8)$$

and from (3.5.7) then also

$$G_q(x) = e^{z(x-1)} \qquad (3.5.9)$$

Thus, for an Erdös-Rényi graph, the two generating functions agree. This is quite useful for deriving analytical results.

When we construct a graph by a stochastic process like that of Erdös-Rényi, the resulting structure need not be connected, but may have several components. In Chap. 2, it was part of the definition of a graph to be connected, but we drop that now because it will complicate our discussion of random graphs. Thus, an Erdös-Rényi graph may have several connected components. In order to understand this better, one lets N tend to ∞ while keeping $z = (N-1)p \sim Np$ from (3.5.1) fixed. It will then depend on the value of that parameter z whether the graph can be expected to contain a giant component or not. Here, a giant component is one that contains a positive fraction of the number of all vertices in the graph. More precisely, when z is above a critical threshold, we expect that our graph contains a component with at least δN vertices, for some $\delta > 0$ that does not depend on N. Below that critical value, all components should have an average size that stays bounded as $N \to \infty$. This fact, and the computation of the value of z where that phase transition occurs, are the basic results of the theory of random graphs. Following [94], we now present a self-consistency argument to derive those results. (For a mathematically more rigorous treatment, we refer to [18].) We already noted that the clustering coefficients will tend to 0 for $N \to \infty$. Therefore, we expect components of bounded size not to contain any triangles. With the same kind of heuristic reasonings, we even assume that all our finite components do not contain cycles, that is, are trees. —Suppose we then randomly choose an edge in our graph, take one of its ends i_0 and look at all the nodes that can be reached via other edges from i_0. Let the number of those nodes be x. We can then consider the generating function $H_1(x)$ for the distribution $p(x)$. When we go from i_0 to any of its new neighbors, that is, other neighbors than that from our edge that we started with, we are in the same situation as before. For any of those neighbors we can again look at the number of nodes that can be reached from it via edges other than that through which we arrived at it from i_0. That number x again is distributed according to the generating function $H_1(x)$. This then directly leads to a self-consistency equation. Namely, the above number k of new neighbors of i_0 is distributed according to q_k from (3.5.4). Also, we recall from (3.1.28) that the generating function for a sum of independent processes is the product of the individual generating functions. Thus, we need to take the product of k factors H_1, one for each new neighbor of i_0 weighted with the probabilities q_k and one additional factor x to account for i_0 itself,[3]

$$H_1(x) = x \sum_{k=0}^{\infty} q_k (H_1(x))^k = x G_q(H_1(x)) \tag{3.5.10}$$

by (3.5.7). From this fixed point equation, we can compute $H_1(x)$. We can then easily determine the distribution $H_0(x)$ for the total size of a finite component. Namely, take any vertex i_1; it has k neighbors with probability p_k, and from each of them, we expect to reach a number x of further vertices distributed according to $H_1(x)$. Therefore, by the same reasoning as before,

$$H_0(x) = x \sum_{k=0}^{\infty} p_k (H_1(x))^k = x G_p(H_1(x)). \tag{3.5.11}$$

All generating functions $G(x) = \sum_k x^k p(k)$ satisfy $G(1) = 1$, see (3.1.25); in particular, this holds for G_p and H_1. Likewise, the expectation value $\langle k \rangle$ of k is given by $G'(1)$, see (3.1.27). Therefore, when all components are finite, the mean component size is

$$\langle x \rangle = H_0'(1) = 1 + G_p'(1) H_1'(1) = 1 + \frac{G_p'(1)}{1 - G_q'(1)} \tag{3.5.12}$$

with the help of (3.5.10). This becomes infinite when $G_q'(1)$ approaches 1, and this then is the phase transition where a giant component appears. From (3.5.7), one can then compute the phase transition value. Above that transition value, the former analysis is no longer valid, but it can nevertheless be used to derive useful results.

---

[3] We have the single node i_0 with probability 1, and so, the generating function is simply $1 \cdot x$.

The point is that while for a giant component we can no longer neglect clustering effects it still applies to the finite components. When σ is the fraction of nodes in the giant component, we then have, using (3.5.11)

$$\sigma = 1 - H_0(1) = 1 - G_p(s), \tag{3.5.13}$$

with $s = H_1(1)$ solving, by (3.5.10),

$$s = G_q(s) \tag{3.5.14}$$

(as in Theorem 3.4.1).

There exist other methods to reduce this percolation analysis to phase transition models in statistical mechanics, and we now discuss one such approach.
A random graph Γ is a member of an ensemble defined by the parameters N and p. Its probability in this ensemble is given by

$$P_\Gamma = p^{l(\Gamma)}(1 - p)^{\binom{N}{2} - l(\Gamma)} = \exp(-\frac{pN^2}{2} + pN(\frac{1}{2}) - \frac{pN}{4} + \frac{l(\Gamma)}{N} + o(1))p^{l(\Gamma)}, \tag{3.5.15}$$

$l(\Gamma)$ denoting the number of edges. The probability of a random graph to have m components then is written as

$$P_m = \sum_\Gamma P_\Gamma \delta(m, m(\Gamma)), \tag{3.5.16}$$

$\delta(m, m(\Gamma))$ being the Kronecker delta. In order to study the distribution of components, one considers

$$P_\Gamma(q) := \frac{1}{z(q)} P_\Gamma q^{m(\Gamma)} \tag{3.5.17}$$

with the normalizing factor

$$z(q) := \sum_\Gamma P_\Gamma q^{m(\Gamma)} = \sum_m P_m q^m. \tag{3.5.18}$$

This quantity is related to the properties of a model from statistical mechanics, the Ising model with q states, also called the Potts model. In this model, one has N spin variables σ_i that can take one of q distinct values $\sigma = 0, 1, \ldots, q - 1$. The energy function of this model (in the so-called mean field variant) is

$$E(\{\sigma_i\}) := -\frac{1}{N} \sum_{i<j} \delta(\sigma_i, \sigma_j) - h \sum_{\sigma=0}^{q-1} f_\sigma \sum_i \delta(\sigma_i, \sigma) \tag{3.5.19}$$

where f_σ is an external field in the spin direction σ, multiplied with the strength h. One introduces a so-called inverse temperature β and encodes the thermodynamic properties of the model at β in the partition function

$$Z_\beta(q) := \sum_{\{\sigma_i\}} \exp(-\beta E(\{\sigma_i\})); \tag{3.5.20}$$

the sum is taken over all possible spin configurations $\{\sigma_i\}$. The partition function can be rewritten as

$$Z_\beta(q) = \sum_{\{\sigma_i\}} \prod_{i<j} (1 + (\exp(\frac{\beta}{N}) - 1)\delta(\sigma_i, \sigma_j)) \exp(\beta h \sum_{\sigma=0}^{q-1} f_\sigma \sum_i \delta(\sigma_i, \sigma)).$$
$$\tag{3.5.21}$$

The relationship of this expression with graphs appears when we expand the product in this expression. Each of the $2^{N(N-1)/2}$ terms corresponds to a graph with N vertices that has an edge between the vertices i and j precisely when they both appear in the corresponding term and when $\sigma_i = \sigma_j$, that is, when the corresponding Kronecker delta takes the value 1. This graph will in general have several components, but the spin values σ_i are constant on each component by our construction. We can then write the partition function as a sum over graphs,

$$Z_\beta(q) = \sum_\Gamma (\exp(\frac{\beta}{N}) - 1)^{l(\Gamma)} \prod_{n=0}^{m(\Gamma)-1} (\sum_\sigma \exp(\beta h f_\sigma S_n)) \tag{3.5.22}$$

where S_n is the size of the nth component and the product extends over the components of Γ. In order to relate this to our ensemble of random graphs, we put $\beta = pN$. Since $\exp(\frac{\beta}{N}) - 1 = \frac{\beta}{N} + O(\frac{1}{N^2})$, we can approximate this for large N as (for $h = 0$)

$$Z_{pN}(q) = \sum_\Gamma p^{l(\Gamma)} q^{m(\Gamma)} = \exp(\frac{pN}{2}) z(q) \tag{3.5.23}$$

in leading order in N, by (3.5.15), (3.5.18). Now, the Potts model exhibits a phase transition to a spontaneous magnetization, that is, all the spins become aligned, above a certain critical value of β. The preceding result then relates this to the appearance of a giant component in our random graph Γ when pN exceeds a critical threshold. The latter is called a percolation phenomenon, and it is thus related to a phase transition in a statistical mechanics model. The parameter h in the latter model is useful for deriving properties by taking derivatives at $h = 0$ whereas the parameter q becomes useful when one studies large deviation properties, that is the properties of atypical members of our ensemble.

We now generalize the construction of Erdös-Rényi by allowing for different connection probabilities for different pairs of vertices. A generalized random graph is characterized by its number N of vertices and real numbers $0 \le p_{ij} \le 1$ (with the symmetry $p_{ij} = p_{ji}$) that assign to each pair i, j of vertices the probability for finding an edge between them. Self-connections of the vertex i are excluded by $p_{ii} = 0$. The expected degree of i then is

$$\nu_i = \sum_j p_{ij}. \tag{3.5.24}$$

A special case which includes scale-free graphs is that of [16]. One starts with an N-tupel $\nu = (\nu_1, ..., \nu_N)$ of positive numbers satisfying

$$\max_i \nu_i^2 \le \sum_j \nu_j; \tag{3.5.25}$$

when the ν_i are positive integers, this is the necessary and sufficient for the existence of a graph with nodes i of degree ν_i, $i = 1, ..., N$. When putting $\gamma := \frac{1}{\sum_i \nu_i}$ and $p_{ij} := \gamma \nu_i \nu_j$, then $0 \le p_{ij} \le 1$ for all i, j. We then insert an edge between the nodes i and j with probability p_{ij} to construct the (generalized) random graph Γ. By (3.5.24), the expected degree of node i in such a graph is ν_i. When all the ν_i are equal, we obtain an Erdös-Rényi graph. For other types, the degree distribution, i.e., the number of nodes i with $\nu_i = k$ will decay as a function of k, at least for large k, for example exponentially. When that number behaves like a power $k^{-\beta}$ instead, we obtain a so-called scale free graph. In the scale-free case, there are thus comparatively more hubs, that is, nodes with large degrees, than in the exponential case. For a systematic treatment of random ensembles of graphs that includes this and other classes, we refer to [77].

Exercises for This Chapter

1. This exercise and the next are about elementary probabilities.—When repeatedly tossing a coin for which heads H turns up with probability q, what is the probability that one needs at least m tosses to see the first H?
2. In this exercise, I shall tell you the first steps of the reasoning and then ask you to continue. The exercise is presented in the guise of a tale taking place in Greek antiquity. At the port of Rhodes, news have just arrived about the results of the competition at Olympeia. An Athenian living at Rhodes therefore goes to the port in order to find out whether his hero Dromeus won the running event. He therefore asks some person idling at the port. However, one third of those people are Cretans, and Cretans always lie. The other two thirds are native Rhodans, and they tell the truth with a probability of three quarters. They do this randomly and independently every time they are asked. The Athenian cannot tell Cretans and Rhodans apart. Thus, when he receives the answer "yes", he reasons as follows. First, there is the probability $\frac{1}{3}$ that the person he has asked is from Crete, and therefore the answer is wrong. In contrast, with probability $\frac{2}{3}$, the person is from Rhodes. In that case, the answer is wrong with probability $\frac{1}{4}$. Thus, altogether, the answer is wrong

with probability $\frac{1}{3} + \frac{2}{3}\frac{1}{4} = \frac{1}{2}$. Thus, the answer is pretty useless (assuming that our Athenian is not a Bayesian with some prior probability different from $\frac{1}{2}$—we do not treat the Bayesian approach here, but if you are so inclined, go ahead and analyze the problem from a Bayesian perspective). However, he does not give up and asks the same person once more, and now receives the answer "no". Thus, he first of all reasons that the person he had asked cannot be a Cretan, as only one of the two answers can be false, but Cretans always give false answers. This does not help much, however, since he cannot tell whether the first or the second answer of the Rhodan he had asked was the correct one. When, in contrast, his second question is again answered with "yes", he reckons that he is either dealing with a Cretan who lied to him twice or with a Rhodan who either told the truth twice or also lied twice. Now the probability that a Rhodan is consistent in that sense is $\frac{3}{4}\frac{3}{4} + \frac{1}{4}\frac{1}{4} = \frac{5}{8}$. Thus, the probability for encountering such a consistent Rhodan is $\frac{2}{3}\frac{5}{8} = \frac{5}{12}$. Thus, the probability to encounter a consistent person is $\frac{5}{12} + \frac{1}{3} = \frac{3}{4}$. Thus, the probability that he is dealing with a Rhodan who twice gave him the correct answer is $\frac{\frac{2}{3}(\frac{3}{4}\frac{3}{4})}{\frac{3}{4}} = \frac{1}{2}$. Thus, the probability that Dromeus won is still $\frac{1}{2}$. So, this still does not really help. Therefore, he asks the same person a third and possibly a fourth time. Can he then infer something useful? And if so, before even asking the first question, how many questions does he expect to have to ask in order to learn the correct result with a probability of at least $\frac{3}{4}$?

3. We now turn to a simple, but more abstract question. Let $X, Y : \Omega \to \mathbb{R}$ be random variables. Show that $Z := \min(X, Y)$ then also is a random variable.

4. Let the random variable C_n denote the result of the nth toss of coin. Assume that

$$p(C_1 = H) = q, \quad p(C_{n+1} = H | C_n = H) = q_1, \quad p(C_{n+1} = H | C_n = T) = q_2$$
$$\text{with} \qquad q_2 < q < q_1 \tag{3.5.26}$$

(that is, having seen H (T) at the nth toss increases the probability of seeing H (T) at the next toss again), and analyze the resulting random process ($n \in \mathbb{N}$).

5. (a) Verify the statements made about the exponential moment generating function $H(t) = E(e^{tX}) = \sum_{n=0}^{\infty} e^{nt} p(n)$ made at the end of Sect. 3.1.

   (b) More generally, let $X = (X^1, \ldots, X^k) : \Omega \to \mathbb{R}^k$ be a tuple of random variables with joint probabilities $p(x^1, \ldots, x^k) = P(X^1 = x^1, \ldots, X^k = x^k)$ for their values. The exponential moment generating functions then is

   $$H(t_1, \ldots, t_k) := E(e^{\sum_{i=1}^{k} t_i X^i}) = \sum_{x^1, \ldots, x^k} p(x^1, \ldots, x^k) e^{\sum_i x^i t_i}.$$
   $$\tag{3.5.27}$$

   Show that H encodes the moments of the distribution p in the sense that the moments are given by

$$E((X^1)^{\nu_1} \cdots (X^k)^{\nu_k}) = \sum_{x^1,\ldots,x^k} \prod_{i=1}^{k} (x^i)^{\nu_i} p(x^1, \ldots, x^k)$$

$$= \frac{\partial^{\sum \nu_i}}{(\partial t_1)^{\nu_1} \cdots (\partial t_k)^{\nu_k}} H(t_1, \ldots, t_k)_{|t_1=0,\ldots,t_k=0}.$$

(c) Use this to verify the formulae for the first and second moments of multinomial sampling in Sect. 3.1 and compute the third and fourth moments.

6. Is the distribution of individuals without descendents in (2.3.17) typical for the statistics of the Wright-Fisher model?

7. We now come to the so-called urn models. We start with Polya's urn. At time 0, this urn is filled with one blue and one red ball. At every positive integer time, one of the balls in the urn is drawn at random and put back together with another ball of the same colour. Thus, at time $n \in \mathbb{N}$, there are $n + 2$ balls in the urn. Let B_n and R_n denote the numbers of blue and red balls at time n. Show that

$$E(B_{n+1}|B_n) = \frac{n+3}{n+2} B_n. \tag{3.5.28}$$

Conclude that $\frac{B_n}{n+2}$ is a martingale. From this, deduce that the ratio $\frac{B_n}{R_n}$ converges almost surely as $n \to \infty$.

8. We now consider an urn that is filled at time 0 with B blue and R red balls. At every integer time, we draw two balls at random. When they have the same colour, we take them away, but if they are of different colours, we return them into the urn. What is the expected number of steps until there is at most one ball of each colour left in the urn?

9. We next come to a problem where I myself do not know the answer, but which probably has either already been solved somewhere in the literature or which can be solved by some bright graduate student. Try to get as far as you can.
   At time 0, we again have an urn with B blue balls and R red balls. In each step, we again randomly draw a pair of balls. Whenever at least one of the balls is red, the pair, that is, both balls, are removed. When both of them are blue, they are both put back into the urn. What is the expected number of steps until no red ball is left in the urn? What is then the expected number of blue balls remaining in the urn? What are the probabilities $P(\mu, \nu)$ for obtaining μ unordered pairs of type (blue, red) and ν pairs of type (red, red) until all the red balls are removed from the urn?
   This problem is inspired by the mating model treated in [106]. The problem can be translated into a mating model as follows. There is a mating pool containing individuals of types A and C. When two individuals of type C encounter each other, nothing happens, and they both remain in the mating pool and go on to look

for new partners, whereas if at least one in a pair is of type A, they mate and are then removed from the mating pool. In other words, individuals of type A mate indiscriminately, whereas individuals of type C only accept partners of type A. (For more on modelling mating systems, see [109].)

Chapter 4
Pattern Formation

Abstract
Questions:

- How will substances diffuse over time?
- How does the biophysics of a neuron work?
- How can we model reaction kinetics in a cell?
- What will happen when two or more species interact, like predators and their prey?
- How can oscillatory patterns emerge?
- How can external stimuli trigger collective behavior within a population of independent individuals?

Understanding pattern formation requires tools from analysis. We introduce dynamical systems to model changes in time and partial differential equations to model distributions in physical or feature spaces. The combination of the two in reaction-diffusion systems leads to mathematical models like the Turing mechanism that can generate surprisingly rich patterns. Another example we treat is chemotaxis where organisms can be induced to collective behavior by following gradients of chemical substances.

We consider spatiotemporal structure formation from interactions between states $f(x, t)$ at points x at times t. This means that the state $f(x_0, t_0)$ is a function of states $f(x, t)$ at some or all other points x at previous times $t \leq t_0$, or at least is influenced by some of those states. Here, space, time, and state space can be discrete or continuous. Discreteness or continuity can lead to rather different effects and difficulties. Perhaps that difference is smallest for space. At the appropriate level of abstraction, discrete and continuous space can be treated in the same manner, although some technical aspects are substantially more difficult in the continuous case. In the discrete case, one assumes some underlying graph structure that incorporates which other points y are the neighbors of a point x_0 whose states $f(y, t)$ then can directly affect the state $f(x_0, t_0)$ for $t_0 \geq t$. In the continuous case, we need a topology on our space that gives us some notion of infinitesimal proximity for setting up partial differential equations as an analytical framework for pattern formation. Concerning time, the discrete case is usually more difficult than the continuous one. In the latter case, we

J. Jost, *Mathematical Methods in Biology and Neurobiology*, Universitext,
DOI: 10.1007/978-1-4471-6353-4_4, © Springer-Verlag London 2014

can work with differential equations whereas in the former one we obtain difference equations or functional iterations. Concerning state space, in the continuous setting we have the possibility of incremental state updates, in particular in the case where time is continuous as well. The discrete case, on the other hand, is more suitable for simulations.

4.1 Partial Differential Equations

4.1.1 The Laplace and the Heat Equation

The theory described in this section and the next is essentially linear and will not by itself have many biological applications. It is rather needed as a preparation for the treatment of reaction-diffusion processes below. As their name indicates, those processes combine linear diffusion processes, as discussed in the present section, with nonlinear reaction dynamics, as treated below. Thereby, they will yield a general and flexible class of models for biological pattern formation.

Partial differential equations, PDEs for short, constitute a field of mathematics that is distinguished from most other mathematical fields by the fact that a definition of its basic object, a partial differential equation, at best is useless and at worst is severely misleading. In order to understand the essence of this field, one rather needs to study prototypical examples, admitting that what constitutes such a prototype is not clearly defined either. Instead of entering into any further generalities, we start with the perhaps most fundamental one, the Laplace equation (although this equation is not directly useful in biology). In this section, we shall only discuss the results, but refer for the proofs to [68] or some other textbook on partial differential equations. For a twice differentiable function $u : \Omega \to \mathbb{R}$ on an open and connected domain $\Omega \subset \mathbb{R}^d$, the Laplacian at $x = (x^1, \ldots, x^d) \in \Omega$ is defined as

$$\Delta u(x) := \sum_{i=1}^{d} \frac{\partial^2 u}{(\partial x^i)^2}(x). \qquad (4.1.1)$$

In the sequel, we shall often abbreviate derivatives by subscripts, i.e.,

$$u_{x^i} := \frac{\partial u}{\partial x^i}, \quad u_{x^i x^i} := \frac{\partial^2 u}{(\partial x^i)^2} \text{ and so on.} \qquad (4.1.2)$$

Thus,

$$\Delta u(x) = \sum_{i=1}^{d} u_{x^i x^i}(x). \qquad (4.1.3)$$

We have already introduced the Laplace operator for a function u on a graph Γ above, in (3),

$$\Delta u(x) := \frac{1}{b_x} \Big(\sum_{y,y\sim x} u(y) - n_x u(x) \Big) \tag{4.1.4}$$

where the vertices y with $y \sim x$ are the neighbors of the vertex x, n_x is the degree of x, that is, the number of its neighbors, and b_x is a positive factor which we preferred to choose as $b_x = n_x$. In order to understand the relationship between these two operators, we replace the domain Ω by its discrete approximation by a grid, as is done for example in numerical schemes for solving PDEs. In order not to have to worry about boundary points, we consider for simplicity the case where Ω is the entire space $\mathbb{R}^d$. For $h > 0$, we then define the discrete space

$$\mathbb{R}_h^d := \{(n_1 h, \ldots, n_d h)\}, n_1, \ldots, n_d \in \mathbb{Z}. \tag{4.1.5}$$

The second partial derivative $u_{x^i x^i}$ then is approximated by the difference

$$u_{ii} := \frac{1}{h^2} (u(x^1, \ldots, x^i + h, \ldots, x^d) + u(x^1, \ldots, x^i - h, \ldots, x^d) \tag{4.1.6}$$
$$- 2u(x^1, \ldots, x^i, \ldots, x^d)),$$

and Δu then is approximated by

$$\Delta_h u := \sum_{i=1}^{d} u_{ii}. \tag{4.1.7}$$

When we consider the grid $\mathbb{R}_h^d$ as a graph on which the neighbors of $x = (x^1, \ldots, x^d)$ are the points $(x^1, \ldots, x^i \pm h, \ldots, x^d), i = 1, \ldots, d$, up to a factor, this is the same as the graph Laplacian (the factor $\frac{1}{h^2}$ has been chosen here in order that the discrete Laplacian converges to the continuous one for $h \to 0$).

Definition 4.1.1 A function u (on a domain Ω or a graph Γ) is called harmonic if it satisfies the Laplace equation

$$\Delta u = 0. \tag{4.1.8}$$

In the discrete case, from (4.1.4) it is clear that a harmonic function u satisfies the mean value property

$$u(x) = \frac{1}{n_x} \sum_{y,y\sim x} u(y) \tag{4.1.9}$$

for all x. The mean value property of harmonic functions also holds in the continuous case:

$$u(x) = \frac{1}{\omega_d r^d} \int_{B(x,r)} u(y) dy \tag{4.1.10}$$

whenever the ball $B(x, r) := \{y \in \mathbb{R}^d : \|y - x\| < r\}$ of radius r around x is contained in Ω. Here, ω_d is the volume of the unit ball in $\mathbb{R}^d$. Conversely, one can show that the mean value property for a continuous u implies that it is harmonic. More generally, u is called subharmonic in Ω if

$$\Delta u(x) \geq 0 \text{ for } x \in \Omega. \tag{4.1.11}$$

This turns out to be equivalent to the mean value inequality

$$u(x) \leq \frac{1}{\omega_d r^d} \int_{B(x,r)} u(y) dy. \tag{4.1.12}$$

From the mean value property, one easily derives the maximum principle:

Lemma 4.1.1 *Suppose that u is harmonic, or more generally, subharmonic in the open and connected Ω. If there exists some $x_0 \in \Omega$ with*

$$u(x_0) = \sup_{x \in \Omega} u(x), \tag{4.1.13}$$

then u is constant in Ω. This is the so-called strong maximum principle, and it implies the weak maximum principle: If Ω is bounded and $u \in C^0(\overline{\Omega})$ (meaning that u is defined and continuous on the closure of Ω), then

$$u(x) \leq \max_{y \in \partial \Omega} u(y). \tag{4.1.14}$$

Finally, if a nonconstant u assumes its maximum at the smooth boundary point y_0 and if it is differentiable there, then

$$\frac{\partial}{\partial n} u(y_0) > 0 \tag{4.1.15}$$

where $\frac{\partial}{\partial n}$ denotes the derivative in the direction of the exterior normal of Ω.

Proof. When $u(x_0) = \sup_{x \in \Omega} u(x) =: M$, we put

$$\Omega^M := \{y \in \Omega : u(y) = M\} \neq \emptyset.$$

For $z \in \Omega^M$ with $r > 0$ such that $B(z, r) \subset \Omega$, we get

$$0 = u(z) - M \leq \frac{1}{\omega_d r^d} \int_{B(z,r)} (u(y) - M)dy \leq 0 \qquad (4.1.16)$$

since M is the supremum of u, and we see that necessarily $u(y) = M$ for all $y \in B(z, r)$. Therefore, whenever $z \in \Omega^M$ and $B(z, r) \subset \Omega$, then that entire ball is also contained in Ω^M. Since Ω is connected, Ω^M has to be all of Ω. This means that $u \equiv M$ in Ω which is what we wanted to prove. The weak maximum principle then follows from the simple observation that a continuous function on the bounded and closed, hence compact set $\overline{\Omega}$ has to assume its supremum. When a harmonic function does so in the interior Ω, it is constant by the strong maximum principle, and (4.1.14) holds, and when the supremum is assumed on the boundary, (4.1.14) holds as well. For the proof of the boundary point maximum result, we refer to the literature, e.g. [68]. $\qquad \square$

The weak maximum principle can also be expressed by saying that a non-constant harmonic function assumes its supremum only on the boundary of Ω when that set is bounded and u is continuous on the closure of Ω.
The strong maximum principle also holds in the discrete case, with the same kind of proof. Since a graph is an object without a boundary, this implies that any harmonic function on a finite graph is constant. Of course, one can also turn the situation into a boundary value problem by declaring a subset S_0 of the vertex set S of Γ as the boundary and considering the problem

$$\Delta u(x) = 0 \text{ for } x \in S\backslash S_0 \qquad (4.1.17)$$
$$u(x) = g(x) \text{ for } x \in S_0$$

for some prescribed function $g : S_0 \to \mathbb{R}$.
By the mean value formula, harmonic functions represent equilibrium states where the value at each point is the average of the values of its neighbors. This observation also suggests a scheme for the proof of the existence of harmonic functions, for example for given boundary values g on $\partial \Omega$. One starts with any (continuous) function $u_0 : \overline{\Omega} \to \mathbb{R}$ with $u_0 = g$ on $\partial\Omega$. Having constructed $u_1, \ldots, u_{n-1}$ iteratively, one finds $u_n(x)$ for $x \in \Omega$ by replacing $u_{n-1}(x)$ by its mean value on some ball $B(x, r) \subset \Omega$. This simple idea can be made to work, and it yields a constructive scheme for finding a harmonic u with given boundary values. On a graph Γ, this means

$$u(x, t + 1) := \frac{1}{n_x} \sum_{y \sim x} u(y, t) \qquad (4.1.18)$$

for $x \in S - S_0, t \in \mathbb{N}$, $u(x, 0)$ being an arbitrary function satisfying the boundary condition (4.1.18). For the numerical implementation (for using this for solving the boundary value problem for harmonic functions in a continuous domain by discrete approximation), one again replaces Ω by a discrete grid of some small mesh size h. For temporal step size k, one puts

$$u(x; t + k) := \frac{1}{2d} \sum_{i=1}^{d} (u(x^1, \ldots, x^i - h, \ldots, x^d; t) \tag{4.1.19}$$
$$+ \, u(x^1, \ldots, x^i + h, \ldots, x^d; t))$$

Conceptually, this is quite useful because it suggests a PDE that is defined on space and time instead of on space only as the Laplace equation and that models the approach to equilibrium. This is the heat equation

$$\frac{\partial}{\partial t} u(x, t) = \Delta u(x, t) \, (= \sum_{i=1}^{d} \frac{\partial^2}{(\partial x^i)^2} u(x, t)) \tag{4.1.20}$$

or abbreviated,

$$u_t(x, t) = \Delta u(x, t). \tag{4.1.21}$$

In fact, the straightforward discretization of (4.1.21) is

$$\frac{1}{k}(u(x, t + k) - u(x, t)) = \frac{1}{h^2} \sum_{i=1}^{d} (u(x^1, \ldots, x^i - h, \ldots, x^d, t) \tag{4.1.22}$$
$$+ \, u(x^1, \ldots, x^i + h, \ldots, x^d, t) - 2u(x^1, \ldots, x^d, t)).$$

For $2dk = h^2$, the term $u(x, t)$ cancels in (4.1.22), and we obtain (4.1.19).[1]
For the heat equation, we also have a maximum principle:

Lemma 4.1.2 *Let $\Omega \subset \mathbb{R}^d$ be open, $0 < T \leq \infty$, and let $u(x, t)$ be continuous for $x \in \overline{\Omega}, 0 \leq t \leq T$ and satisfy*

$$u_t(x, t) = \Delta u(x, t) \text{ for } x \in \Omega, 0 < t < T. \tag{4.1.23}$$

Then

$$\sup_{\overline{\Omega} \times [0,T]} u = \sup_{(\overline{\Omega} \times \{0\}) \cup (\partial \Omega \times [0,T])} u. \tag{4.1.24}$$

When $T < \infty$, the supremum becomes a maximum. Again, there is also a strong version of the maximum principle, saying that a solution of (4.1.23) cannot attain a maximum in $\Omega \times (0, T]$ without being constant. Also, there is an analogue of the result for boundary maxima, that is, at a nontrivial boundary point maximum, one obtains a positive exterior normal derivative, if the situation is sufficiently smooth.

---

[1] The fact that the temporal step size k satisfies $2dk = h^2$ slows down the convergence of the scheme for $h \to 0$ and makes this not really a good numerical scheme in practice.

The Lemma says that the maximum of a solution of the heat equation is always attained either at the spatial boundary $\partial\Omega \times [0, T]$ of the cylinder $\Omega \times (0, T)$ or at the initial set $t = 0$.

In the continuous as in the discrete case, one shows that a solution of the initial boundary value for the heat equation

$$u_t(x, t) = \Delta u(x, t) \text{ for } x \in \Omega, 0 < t \tag{4.1.25}$$

$$u(x, 0) = u_0(x) \text{ for } x \in \Omega \tag{4.1.26}$$

$$u(y, t) = g(y) \text{ for } y \in \partial\Omega \tag{4.1.27}$$

for continuous initial values u_0 and boundary values g (and under certain mild technical assumptions on Ω) converges to a solution of the boundary value problem for the Laplace equation, the Dirichlet problem

$$\Delta u(x) = 0 \text{ for } x \in \Omega \tag{4.1.28}$$

$$u(y) = g(y) \text{ for } y \in \partial\Omega$$

for $t \to \infty$, that is, $\lim_{t\to\infty} u(x, t) = u(x)$.

A perhaps simpler boundary condition is the periodic boundary condition. We consider a domain of the form $\Omega = (0, L_1)\times, \ldots, \times(0, L_d) \subset \mathbb{R}^d$ and require for $u : \bar{\Omega} \to \mathbb{R}$ that

$$\Delta u(x) = 0 \text{ for } x \in \Omega \tag{4.1.29}$$

$$u(x_1, \ldots, x_{i-1}, L_i, x_{i+1}, \ldots, x_d) = u(x_1, \ldots, x_{i-1}, 0, x_{i+1}, \ldots, x_d)$$

for all $x = (x_1, \ldots, x_d) \in \Omega$, $i = 1, \ldots, d$. This means that u can be periodically extended from Ω to all of $\mathbb{R}^d$. Of course, this boundary condition for is less general than (4.1.28) because it can be posed only on rectangular domains. In contrast to (4.1.28), this problem is trivial, and the only solutions are the constants.

A generalization of the Laplace equation is the Poisson equation

$$\Delta u(x) = f(x) \text{ for } x \in \Omega \tag{4.1.30}$$

for some given function $f : \Omega \to \mathbb{R}$, or its analogues in the discrete case or for the heat equation. Again, we can impose boundary conditions of Dirichlet type or periodic ones. Here, the periodic boundary value problem is no longer trivial. In contrast to the Dirichlet boundary value problem, however, the solution of the periodic boundary value problem for the Poisson equation is not unique; it is determined only up to an additive constant.

A basic idea for solving (4.1.30) consists in the superposition of point solutions. That means that for each $y \in \Omega$, we try to find some function $\gamma(x, y)$ that solves (4.1.30) at y and is harmonic elsewhere. If we then integrate w.r.t. y, we should obtain the desired solution of (4.1.30). Let us first try to implement this in the discrete case where we only have to take a sum in place of an integral. Of course, instead of $f(y)$, we can then take 1 as the right hand side for our equation at y and multiply the result by $f(y)$ and then sum w.r.t y. Thus, given a graph Γ as before, and a node $y \in S$, the vertex set of Γ, we want to solve

$$\Delta_x G(x, y) \left(= \frac{1}{n_x} \sum_{z, z \sim x} (G(z, y) - G(x, y)) \right) = \frac{1}{n_x} \delta(x, y) \qquad (4.1.31)$$

with

$$\delta(x, y) := \begin{cases} 1 \text{ for } x = y \\ 0 \text{ elsewhere.} \end{cases} \qquad (4.1.32)$$

(Since we are looking at functions of two variables, we indicate by a subscript w.r.t. which variable the Laplacian Δ acts.) A solution of (4.1.31) is called a Green function. If we can find such a Green function, a solution to the discrete Poisson equation

$$\Delta u(x) = f(x) \text{ for } x \in S \qquad (4.1.33)$$

is then simply given by

$$u(x) = n_x \sum_y G(x, y) f(y). \qquad (4.1.34)$$

There is one problem here: We cannot solve (4.1.31) because for any function g on a graph Γ, we have

$$\sum_x n_x \Delta g(x) = 0 \qquad (4.1.35)$$

and therefore necessarily also

$$\sum_x n_x \Delta_x G(x, y) = 0, \qquad (4.1.36)$$

but the right hand side of (4.1.31) does not fulfill that condition. In abstract terms, the Laplacian has a kernel, consisting of the constant functions, and is therefore not invertible. It is invertible only on the space orthogonal to the constants. That means that we can expect to solve (4.1.33) only when f satisfies

$$\sum_x n_x f(x) = 0. \qquad (4.1.37)$$

This can be easily remedied, however. We simply replace (4.1.31) by

$$\Delta_x G(x, y) = \frac{1}{n_x}\delta(x, y) - \frac{1}{\sum_z n_z}, \tag{4.1.38}$$

that is, subtract a suitable constant on the right hand side so as to achieve (4.1.36). When (4.1.37) holds, the contribution of the constant disappears in (4.1.34), and so, we can solve (4.1.33) for those f.

Another possibility to circumvent that problem is to impose a boundary condition, that is, solve

$$\Delta u(x) = f(x) \text{ for } x \in S\backslash S_0 \tag{4.1.39}$$

$$u(x) = g(x) \text{ for } x \in S_0 \tag{4.1.40}$$

for some prescribed function $g : S_0 \to \mathbb{R}$. Here, we assume $S_0 \neq \emptyset$, but otherwise, S_0 is completely arbitrary.

In order to achieve that, we first consider the homogeneous boundary condition, that is, $g = 0$. For that, we impose the homogeneous boundary condition

$$G(x, y) = 0 \text{ for } x \in S_0 \text{ and all } y \tag{4.1.41}$$

take the corresponding u from (4.1.34) (the equation now imposed for $x \in S\backslash S_0$), which then satisfies $u(x) = 0$ for $x \in S_0$. In order to solve the general boundary value problem, we then simply add a solution $u_0(x)$ of (4.1.17) to get the right boundary condition. In abstract terms, imposing a boundary condition eliminates the kernel of the Laplacian. We can then not only solve the boundary value problem, but the solution is also unique, because the difference of two solutions is a harmonic function with zero boundary values, hence identically zero itself (as follows in many ways, for example from the maximum principle).

In the continuous case, we can use the same strategy. We want to solve

$$\Delta u(x) = f(x) \text{ for } x \in \Omega \tag{4.1.42}$$
$$u(x) = g(x) \text{ for } x \in \partial\Omega. \tag{4.1.43}$$

Again, assuming that we can already solve the boundary value problem for the Laplace equation, that is, find a solution for

$$\Delta u(x) = 0 \text{ for } x \in \Omega \tag{4.1.44}$$
$$u(x) = g(x) \text{ for } x \in \partial\Omega, \tag{4.1.45}$$

we consider homogenous boundary values, that is,

$$\Delta u(x) = f(x) \text{ for } x \in \Omega \tag{4.1.46}$$
$$u(x) = 0 \text{ for } x \in \partial\Omega. \tag{4.1.47}$$

As before, we start with

$$\Delta_x G(x, y) = \delta(x, y), \tag{4.1.48}$$

the Dirac delta functional. This means that for every continuous ϕ, we have

$$\phi(x) = \int_\Omega \delta(x, y)\phi(y)dy = \int_\Omega \Delta_x G(x, y)\phi(y)dy. \tag{4.1.49}$$

And

$$u(x) = \int_\Omega G(x, y)f(y)dy \tag{4.1.50}$$

then satisfies

$$\Delta u(x) = \int_\Omega \Delta_x G(x, y)f(y)dy = f(x). \tag{4.1.51}$$

Once more, in order to get homogeneous boundary values, that is, $u_{|\partial\Omega} = 0$, for u in (4.1.50), we need to have $G(x, y) = 0$ for $x \in \partial\Omega$. This can indeed be achieved, but we do not go into the details here. We rather display the so-called fundamental solutions, particular solutions of (4.1.48) in the whole space $\mathbb{R}^d$. These are

$$\Gamma(x, y) = \Gamma(|x - y|) := \begin{cases} \frac{1}{2\pi} \log |x - y| & \text{for } d = 2 \\ \frac{1}{d(2-d)\omega_d} |x - y|^{2-d} & \text{for } d > 2 \end{cases} \tag{4.1.52}$$

where ω_d is the volume of the d-dimensional unit ball $B(0, 1) \subset \mathbb{R}^d$. The computations that this Γ solves (4.1.48) are straightforward, but somewhat lengthy and omitted here.

For the heat equation, we also have a fundamental solution from which more general problems can be solved by superposition. For $x, y \in \mathbb{R}^d, t > 0$, we put

$$K(x, y, t) := \frac{1}{(4\pi t)^{d/2}} e^{-\frac{|x-y|^2}{4t}}. \tag{4.1.53}$$

K solves the heat equation:

$$\frac{\partial}{\partial t} K(x, y, t) = \Delta_x K(x, y, t) \text{ for all } x, y \in \mathbb{R}^d, t > 0. \tag{4.1.54}$$

We have the normalization

$$\int_{\mathbb{R}^d} K(x, y, t)dy = 1 \text{ for all } x \in \mathbb{R}^d, t > 0. \tag{4.1.55}$$

Also, for a bounded and continuous function f on $\mathbb{R}^d$,

$$u(x, t) = \int_{\mathbb{R}^d} K(x, y, t) f(y) dy \qquad (4.1.56)$$

solves the heat equation

$$u_t = \Delta u \qquad (4.1.57)$$

for $x \in \mathbb{R}^d$, $t > 0$ and has the initial values

$$\lim_{t \to 0} u(x, t) = f(x) \qquad (4.1.58)$$

which is abbreviated as

$$u(x, 0) = f(x). \qquad (4.1.59)$$

4.1.2 The Eigenvalue Problem for the Laplace Operator and Expansions of Solutions of PDEs in Terms of Eigenfunctions

We now briefly discuss the eigenvalue problem for the Laplace operator and its connections with the heat equation. Again, this is formally analogous to the discrete case, already treated in Sect. 2.2.3, although the details now require a more careful analysis and depend on some analytical result, the Rellich compactness theorem. The eigenvalue problem for the Laplace operator consists in finding nontrivial solutions of

$$\Delta u(x) + \lambda u(x) = 0 \quad \text{in } \Omega, \qquad (4.1.60)$$

for some constant λ, the eigenvalue in question. Here one also imposes some boundary conditions on u. It seems natural to require the Dirichlet boundary condition

$$u = 0 \quad \text{on } \partial\Omega. \qquad (4.1.61)$$

For many applications, however, it is more natural to have the Neumann boundary condition

$$\frac{\partial u}{\partial n} = 0 \quad \text{on } \partial\Omega \qquad (4.1.62)$$

instead, where $\frac{\partial}{\partial n}$ denotes the derivative in the direction of the exterior normal. Here, in order to make this meaningful, one needs to impose suitable regularity of $\partial\Omega$. For simplicity, we shall assume that Ω is a C^∞-domain in treating Neumann boundary conditions. For suitable domains, we can also impose periodic boundary conditions, as discussed above, see (4.1.29). When the domain is a closed manifold (compact, without boundary), e.g., a torus, in place of a subset with boundary of $\mathbb{R}^d$, one does not impose any further condition, as in the case of a graph.

We shall employ the L^2-product

$$\langle f, g \rangle := \int_\Omega f(x)g(x)dx \tag{4.1.63}$$

for $f, g \in L^2(\Omega)$, that is, $\int_\Omega f(x)^2 dx$, $\int_\Omega g(x)^2 dx < \infty$ and we shall also put

$$\|f\| := \|f\|_{L^2(\Omega)} = \langle f, f \rangle^{\frac{1}{2}}. \tag{4.1.64}$$

We note the symmetry of the Laplace operator,

$$\langle \Delta\varphi, \psi \rangle = -\langle D\varphi, D\psi \rangle = \langle \varphi, \Delta\psi \rangle \tag{4.1.65}$$

for all $\varphi, \psi \in C_0^\infty(\Omega)$, as well as for $\varphi, \psi \in C^\infty(\Omega)$ with $\frac{\partial\varphi}{\partial n} = 0 = \frac{\partial\psi}{\partial n}$ on $\partial\Omega$. Here, $D\varphi$ abbreviates the vector $\frac{\partial}{\partial x^1}\varphi, \ldots, \frac{\partial}{\partial x^d}\varphi$.

This symmetry implies that all eigenvalues are real.

Theorem 4.1.1 *Let $\Omega \subset \mathbb{R}^d$ be connected, open and bounded. Then the eigenvalue problem*

$$\Delta u + \lambda u = 0, \quad u = 0 \text{ on } \partial\Omega$$

has countably many eigenvalues

$$0 < \lambda_1 < \lambda_2 \leq \cdots \leq \lambda_m \leq \cdots \tag{4.1.66}$$

with

$$\lim_{m \to \infty} \lambda_m = \infty$$

and pairwise L^2-orthonormal eigenfunctions u_i and $\langle Du_i, Du_i \rangle = \lambda_i$. Any $v \in L^2(\Omega)$ can be expanded in terms of these eigenfunctions,

$$v = \sum_{i=1}^\infty \langle v, u_i \rangle u_i \quad \text{(and thus } \langle v, v \rangle = \sum_{i=1}^\infty \langle v, u_i \rangle^2). \tag{4.1.67}$$

Moreover, the first eigenfunction u_1 does not change sign in Ω, that is, we may assume

$$u_1 > 0 \text{ in } \Omega. \tag{4.1.68}$$

We should explain the inequality signs in (4.1.66). An eigenvalue can have higher multiplicity, meaning that there may exist several linearly independent eigenfunctions with the same eigenvalue. Therefore, eigenvalues are counted according to the dimension of the eigenspaces. The inequality $0 < \lambda_1$ simply says that the first eigenvalue is positive. The inequality $\lambda_1 < \lambda_2$ represents the theorem that the first eigenvalue is simple, that is, the eigenspace corresponding to λ_1 is one-dimensional. This fact depends on the assumption that Ω is connected.

For Neumann boundary conditions, we have an analogous result:

Theorem 4.1.2 *Let $\Omega \subset \mathbb{R}^d$ be bounded, open, and of class C^∞. Then the eigenvalue problem*

$$\Delta u + \lambda u = 0, \quad \frac{\partial u_i}{\partial n} = 0 \ \ on \ \partial\Omega$$

has countably many eigenvalues

$$0 = \lambda_0 \leq \lambda_1 \leq \cdots \leq \lambda_m \leq \cdots$$

with

$$\lim_{n \to \infty} \lambda_m = \infty$$

and pairwise L^2-orthonormal eigenfunctions u_i. Any $v \in L^2(\Omega)$ can be expanded in terms of these eigenfunctions

$$v = \sum_{i=0}^{\infty} \langle v, u_i \rangle u_i \quad (and \ thus \ \langle v, v \rangle = \sum_{i=0}^{\infty} \langle v, u_i \rangle^2). \tag{4.1.69}$$

In Theorem 4.1.2, $\lambda_0 = 0$ appears as an eigenvalue. In fact, any non-vanishing constant is an eigenfunction with eigenvalue 0, and, in contrast to the Dirichlet condition, these are not excluded by the Neumann boundary condition. When Ω has more than one component, we can in fact choose a different constant on each component. When Ω is connected, however, a global constant is the only eigenfunction with eigenvalue 0, and this then is a simple eigenvalue.

It is also insightful and instructive to look at the scaling behavior of the eigenvalues. If instead of Ω we consider the domain $\alpha\Omega := \{\alpha x : x \in \Omega\}$ for a scaling factor $\alpha > 0$, then its eigenfunctions are given by $u_i(\frac{y}{\alpha})$. Since $\frac{\partial^2}{(\partial y^i)^2} u(\frac{y}{\alpha}) = \frac{\partial^2}{(\partial x^i)^2} u(x)$ for $y^i = \alpha x^i$, the eigenvalues of Ω_α then are $\alpha^{-2}\lambda_i$ where the λ_i, of course, are those of Ω (this argument is valid for both Dirichlet and Neumann eigenvalues). Since the volume $\|\Omega_\alpha\|$ of Ω_α is $\alpha^d \|\Omega\|$ for a d-dimensional domain, the eigenvalues scale like $Vol^{\frac{-2}{d}}$. The Weyl type estimates state that (under some mild regularity assumptions on Ω,) the eigenvalues λ_k of Ω grow proportionally to $(\frac{k}{|\Omega|})^{\frac{2}{d}}$ up to terms of lower order.

Below, we shall also need the following result. We consider the average of v on Ω

$$\bar{v} := \frac{1}{\|\Omega\|} \int_{\Omega} v(x)dx \tag{4.1.70}$$

where $\|\Omega\|$ is the volume of Ω.

Corollary 4.1.1 *Let λ_1 be the first nontrivial Neumann eigenvalue of Ω. For $v \in H^{1,2}(\Omega)$ (that is, it not only is square integrable itself, but also has square integrable first derivatives in the L^2-sense) with $\frac{\partial v}{\partial \nu}$ on $\partial\Omega$*

$$\lambda_1 \langle v - \bar{v}, v - \bar{v} \rangle \le \langle Dv, Dv \rangle. \tag{4.1.71}$$

For v of class $H^{2,2}(\Omega)$ (that is, it also has square integrable second derivatives in the L^2-sense), also

$$\lambda_1 \langle Dv, Dv \rangle \le \langle \Delta v, \Delta v \rangle. \tag{4.1.72}$$

Proof. We observe that $Dv = D(v - \bar{v})$, $\Delta v = \Delta(v - \bar{v})$, and

$$\langle v - \bar{v}, v - \bar{v} \rangle = \sum_{i=1}^{\infty} \langle v, u_i \rangle^2, \tag{4.1.73}$$

that is, the term for $i = 0$ disappears from the expansion because $v - \bar{v}$ is orthogonal to the constant eigenfunction u_0. Using

$$\langle Dv, Dv \rangle = \sum_{i=1}^{\infty} \lambda_i \langle v, u_i \rangle^2$$

$$\langle \Delta v, \Delta v \rangle = \sum_{i=1}^{\infty} \lambda_i^2 \langle v, u_i \rangle^2$$

and $\lambda_1 \le \lambda_i$ then yields (4.1.71), (4.1.72). $\square$

Remark: The following argument that assumes still more regularity of v is also instructive:

$$\int_{\Omega} (\Delta v)^2 = \int \sum_{i=1}^{d} v_{x^i x^i} \sum_{j=1}^{d} v_{x^j x^j} = \int \sum_{i,j=1}^{d} v_{x^i x^j} v_{x^i x^j}, \tag{4.1.74}$$

integrating by parts twice, without incurring a boundary term because of the Neumann boundary condition. Therefore, applying (4.1.71) to v_{x^i} for $i = 1, \ldots, d$ yields (4.1.72).

One can use the eigenfunctions of the Laplacian to write an expansion for the Green function. We consider the case of the Dirichlet boundary conditions as in

Theorem 4.1.1. Thus, the Green function has to solve

$$\Delta_x G(x, y) = \delta(x, y) \text{ for } x, y \in \Omega \tag{4.1.75}$$

$$G(x, y) = 0 \text{ for } x \in \partial\Omega, y \in \Omega. \tag{4.1.76}$$

This Green function can then be represented in terms of the Dirichlet eigenfunctions of Theorem 4.1.1 as

$$G(x, y) = -\sum_n \frac{1}{\lambda_n} u_n(x) u_n(y). \tag{4.1.77}$$

To see this, recalling (4.1.50), (4.1.51), for $f \in L^2(\Omega)$, we consider

$$u(x) = -\int_\Omega \sum_n \frac{1}{\lambda_n} u_n(x) u_n(y) f(y) dy \tag{4.1.78}$$

and compute

$$\begin{aligned}
\Delta u(x) &= \sum_n u_n(x) \int u_n(y) f(y) dy \\
&= \sum_n u_n(x) \langle u_n, f \rangle \\
&= f(x) \text{ by (4.1.67).}
\end{aligned}$$

These expansions in terms of eigenfunctions of the Laplace operator are also useful for the heat equation

$$u_t(x, t) = \Delta u(x, t) \quad \text{for } x \in \Omega, 0 < t. \tag{4.1.79}$$

We try to find solutions with separated variables, i.e., of the form

$$u(x, t) = v(x) w(t). \tag{4.1.80}$$

Inserting this ansatz into (4.1.79), we obtain

$$\frac{w_t(t)}{w(t)} = \frac{\Delta v(x)}{v(x)}. \tag{4.1.81}$$

Since the left-hand side of (4.1.81) is a function of t only, while the right-hand side is a function of x, each of them has to be constant. Thus

$$\Delta v(x) = -\lambda v(x), \tag{4.1.82}$$

$$w_t(t) = -\lambda w(t), \tag{4.1.83}$$

for some constant λ. We consider the case where we assume homogeneous boundary conditions on $\partial\Omega \times [0, \infty)$, i.e.,

$$u(x, t) = 0 \quad \text{for } x \in \partial\Omega \tag{4.1.84}$$

or equivalently,

$$v(x) = 0 \quad \text{for } x \in \partial\Omega. \tag{4.1.85}$$

A nontrivial solution v of (4.1.82), (4.1.85) is an eigenfunction of the Laplace operator, and λ an eigenvalue. By Theorem 4.1.1, the eigenvalues constitute a discrete sequence $(\lambda_n)_{n\in\mathbb{N}}$, $\lambda_n \to \infty$ for $n \to \infty$. Thus, a nontrivial solution of (4.1.82), (4.1.85) exists precisely if $\lambda = \lambda_n$, for some $n \in \mathbb{N}$. The solution of (4.1.83) then is simply given by

$$w(t) = w(0)e^{-\lambda t}.$$

So, if we denote an eigenfunction for the eigenvalue λ_n by u_n, we obtain the solution

$$u(x, t) = u_n(x)w(0)e^{-\lambda_n t}$$

of the heat Eq. (4.1.79), with the homogeneous boundary condition

$$u(x, t) = 0 \quad \text{for } x \in \partial\Omega$$

and the initial condition

$$u(x, 0) = u_n(x)w(0).$$

This seems to be a rather special solution. Nevertheless, in a certain sense this is the prototype of a solution. As already noted above, we have a superposition principle. Since (4.1.79) is a linear equation, any linear combination of solutions is a solution itself, and so we may take sums of such solutions for different eigenvalues λ_n. In fact, by Theorem 4.1.1, any L^2-function on Ω, and thus in particular any continuous function f on $\overline{\Omega}$, assuming Ω to be bounded, that vanishes on $\partial\Omega$, can be expanded as

$$f(x) = \sum_{n\in\mathbb{N}} \alpha_n u_n(x). \tag{4.1.86}$$

Here, the $u_n(x)$ are the orthonormal eigenfunctions of Δ,

$$\int_\Omega u_n(x)u_m(x)dx = \delta_{nm},$$

and with

$$\alpha_n = \int_\Omega u_n(x) f(x) dx.$$

We then have an expansion for the solution of

$$
\begin{aligned}
u_t(x, t) &= \Delta u(x, t) && \text{for } x \in \Omega, t \geq 0, \\
u(x, t) &= 0 && \text{for } x \in \partial\Omega, t \geq 0, \\
u(x, 0) &= f(x) && \left(= \sum_n \alpha_n u_n(x) \right), && \text{for } x \in \Omega,
\end{aligned}
\tag{4.1.87}
$$

namely,

$$u(x, t) = \sum_{n \in \mathbb{N}} \alpha_n e^{-\lambda_n t} u_n(x). \tag{4.1.88}$$

Since all the λ_n are nonnegative, we see from this representation that all the "modes" $\alpha_n u_n(x)$ of the initial values f are decaying in time for a solution of the heat equation. In this sense, the heat equation regularizes or smoothes out its initial values. In particular, since thus all factors $e^{-\lambda_n t}$ are less than or equal to 1 for $t \geq 0$, the series (4.1.88) converges in $L^2(\Omega)$, because (4.1.86) does.
If we write

$$q(x, y, t) := \sum_{n \in \mathbb{N}} e^{-\lambda_n t} u_n(x) u_n(y), \tag{4.1.89}$$

Theorem 4.1.1 shows convergence of this series, and we may represent the solution $u(x, t)$ of (4.1.87) as

$$
\begin{aligned}
u(x, t) &= \sum_{n \in \mathbb{N}} e^{-\lambda_n t} u_n(x) \int_\Omega u_n(y) f(y) dy && \text{by (4.1.88)} \\
&= \int_\Omega q(x, y, t) f(y) dy.
\end{aligned}
\tag{4.1.90}
$$

Comparing this with (4.1.56), (4.1.57), we see that $q(x, y, t)$ as in (4.1.89) yields a heat kernel, in analogy to formula (4.1.78) for the Laplace equation.

4.2 Diffusion and Random Walks

4.2.1 Random Walks on Graphs

In this section, we want to explore the relationship between partial differential equations and stochastic analysis. As before, we start with the discrete case. In the classical treatment of [30], this was carried for lattices in $\mathbb{R}^d$, for the purpose of discretizing linear partial differential equations. Here, we take a general graph Γ with vertex set V. In contrast to [30], we shall have to work with two different operators, depending on whether we vary the starting or the target point of a random walk. We also take some subset V_0 of V as the boundary. $V \setminus V_0$ then is the interior. The choice of V_0 is rather arbitrary. It should not be empty, at least on certain occasions, nor should it coincide with the entire vertex set. Although we shall not indicate those places below where such assumptions are used, we might also wish to require that the graph obtained from Γ by eliminating the vertex set V_0 and all edges connected to vertices in V_0 is still connected.

We now construct a diffusion process on Γ. We assume that we have a unit of some substance at the point $x \in V$. That substance is diffusing in Γ in such manner that the fraction s of our substance present at the vertex y at time n is equally distributed among the neighbors of y at time $n + 1$, that is, each neighbor of y receives $\frac{s}{n_y}$ of the substance where n_y, as always, is the degree of y. Whatever amount of the substance reaches a boundary point will stay there forever. When the initial point x was a boundary point, the whole amount of the substance will stay there. Thus, the boundary is absorbing for our diffusion process.—According to these rules, the total amount of our substance present in Γ at any time n is always the same, that is, our diffusion process satisfies a conservation law.

There is an alternative view of this process, and identifying those two views will be very insightful. That latter view considers a random walk on Γ. This is a stochastic process, discrete Brownian motion, with discrete time $n \in \mathbb{N}$. A walker or a Brownian particle, whatever physical interpretation one prefers, starts at x, and when it happens to be at the interior point y at time n, it moves to one of the neighbors of y at time $n + 1$, and all these neighbors have the same probability, that is $\frac{1}{n_y}$, of receiving the particle.

We see that the probabilities follow the above diffusion process. Let z be a boundary point. The probability $w(x, z)$ of reaching z by a random walk starting at x without having previously hit any other boundary point then equals the fraction of our diffusing substance that has accumulated at z in infinite time. According to our rules at the boundary, we also have

$$ w(z, z) = 1 \text{ for } z \in V_0 \quad \text{and } w(z, z') = 0 \text{ when } z \neq z' \in V_0. \tag{4.2.1} $$

When z is the first boundary point reached by a random walk starting at x, we also call z the exit point of that random walk. When $r_n(x, z)$ is the fraction of our substance reaching z from x after precisely n steps, we have

$$ w(x, z) = \sum_{\nu=0}^{\infty} r_\nu(x, z). \tag{4.2.2} $$

Likewise, the probability $q_n(x, y)$ of reaching y after n steps starting from x equals the fraction of our substance that happens to be at time n at y, in case x and y are interior points. When either of them is a boundary point, that probability is put to 0. We are also interested in the sum

$$v(x, y) := \sum_{\nu=0}^{\infty} q_\nu(x, y). \tag{4.2.3}$$

When x and y are interior points, this equals the amount of substance that has passed through y at some time. In the probabilistic interpretation, this is the expected number of times the random walk starting at x passes through y before exiting at some boundary point.

The sum in (4.2.2) converges because its members are nonnegative and its partial sums cannot become larger than 1 as only some fraction of the original substance can reach z before being absorbed at some other boundary point. This then also implies that $q_n(x, y)$ tends to 0 for $n \to \infty$. Indeed, let $q_n(x, y) > \epsilon$, and assume that the boundary point z can be reached from y along some path $y_0 := y, y_1, \ldots, y_m = z$ that does not hit the boundary before z. Then after m steps, a fraction $\dfrac{\epsilon}{n_{y_0} n_{y_1} \cdots n_{y_{m-1}}}$ of our substance reaches z along that particular path and is absorbed at z. By convergence of the series in (4.2.2), the fraction of substance reaching z after n steps tends to 0 for $n \to \infty$. Therefore, $q_n(x, y)$ also has to converge to 0 for $n \to \infty$. In particular, the probability of staying in the interior for an infinite amount of time vanishes. Consequently,

$$\sum_{z \in V_0} w(x, z) = 1. \tag{4.2.4}$$

We may then consider $w(x, .)$ as a probability distribution for the exit point of the random walk starting at x.

We now turn to proving the convergence of the series in (4.2.3). We have the relation

$$q_n(x, y) = \sum_{y' \sim y} \frac{1}{n_{y'}} q_{n-1}(x, y') \tag{4.2.5}$$

for $n > 1$ because whatever is reaching y at some time n has to be at some neighbor y' of y at time $n - 1$. We recall here that $q_{n-1}(x, y') = 0$ when y' happens to be a boundary point. Also, $q_0(x, x) = 1$ when x itself is an interior point, $=0$ when it is a boundary point.

The same type of reasoning also yields a more general relation,

$$q_n(x, y) = \sum_{z \in V \setminus V_0} q_{n_1}(x, z) q_{n-n_1}(z, y) \text{ whenever } n_1 < n \tag{4.2.6}$$

where the sum now extends over all interior vertices. This relation follows from the simple observation that whatever reaches y from x in n steps has to be at some interior vertex at the time $n_1 < n$ whence it arrives at y after $n - n_1$ further steps. Obviously, (4.2.5) is a special case of (4.2.6), corresponding to $n_1 = n - 1$. Returning to (4.2.5), we see that the partial sums

$$v_n(x, y) := \sum_{\nu=0}^{n} q_\nu(x, y) \tag{4.2.7}$$

satisfy

$$v_n(x, y) = \begin{cases} \sum_{y' \sim y} \frac{1}{n_{y'}} v_{n-1}(x, y') & \text{for } x \neq y \\ 1 + \sum_{y' \sim y} \frac{1}{n_{y'}} v_{n-1}(x, y') & \text{for } x = y \end{cases} \tag{4.2.8}$$

We define the operator Δ' by

$$\Delta' f(y) := \sum_{y' \sim y} \frac{1}{n_{y'}} f(y') - f(y). \tag{4.2.9}$$

When our graph is regular in the sense that all vertices y have the same degree $n_y = k$, say, then $\Delta' = \Delta$. For other graphs, the two operators are obviously different. From (4.2.5), we infer

$$q_{n+1}(x, y) - q_n(x, y) = \Delta'_y q_n(x, y) \tag{4.2.10}$$

and $q_n(x, y) = 0$ when y is a boundary point. This is a discrete heat equation, with time step 1 and Δ' in place of Δ.
Similarly, from (4.2.8), (4.2.7), we infer

$$\Delta'_y v_n(x, y) = \begin{cases} q_{n+1}(x, y) & \text{for } x \neq y \\ q_{n+1}(x, y) - 1 & \text{for } x = y \end{cases} \tag{4.2.11}$$

and $v_n(x, y) = 0$ when y is a boundary point. Again, we can rewrite this is as a heat type equation

$$\Delta'_y v_n(x, y) = \begin{cases} v_{n+1}(x, y) - v_n(x, y) & \text{for } x \neq y \\ v_{n+1}(x, y) - v_n(x, y) - 1 & \text{for } x = y \end{cases} \tag{4.2.12}$$

Since we already know that $q_n(x, y) \to 0$ for $n \to \infty$, $v_n(x, y)$ converges to the solution v of

$$\Delta'_y v(x, y) = \begin{cases} 0 & \text{for } x \neq y \\ -1 & \text{for } x = y \end{cases} \tag{4.2.13}$$

with boundary values 0.
When we vary x in place of y, we come up with the Laplacian Δ in place of Δ'.
Indeed, we have

$$q_n(x, y) = \frac{1}{n_x} \sum_{x' \sim x} q_{n-1}(x', y) \qquad (4.2.14)$$

because any random path that goes from x to y in n steps has to pass through one of
the neighbors of x in the first step with equal probability $\frac{1}{n_x}$. (This is again a special
case of (4.2.6), this time for $n_1 = 1$.) Therefore, we obtain the discrete heat equation

$$q_{n+1}(x, y) - q_n(x, y) = \Delta_x q_n(x, y). \qquad (4.2.15)$$

As before, from (4.2.14), we conclude

$$v_n(x, y) = \begin{cases} \frac{1}{n_x} \sum_{x' \sim x} v_{n-1}(x', y) & \text{for } x \neq y \\ 1 + \frac{1}{n_x} \sum_{x' \sim x} v_{n-1}(x', y) & \text{for } x = y \end{cases} \qquad (4.2.16)$$

From (4.2.8), (4.2.16), we infer

$$\Delta_x v_n(x, y) = \begin{cases} q_{n+1}(x, y) & \text{for } x \neq y \\ q_{n+1}(x, y) - 1 & \text{for } x = y \end{cases} \qquad (4.2.17)$$

and $v_n(x, y) = 0$ when x is a boundary point or, equivalently,

$$\Delta_x v_n(x, y) = \begin{cases} v_{n+1}(x, y) - v_n(x, y) & \text{for } x \neq y \\ v_{n+1}(x, y) - v_n(x, y) - 1 & \text{for } x = y. \end{cases} \qquad (4.2.18)$$

Since we already know that $q_n(x, y) \to 0$ for $n \to \infty$, $v_n(x, y)$ converges to the
solution v of

$$\Delta_x v(x, y) = \begin{cases} 0 & \text{for } x \neq y \\ -1 & \text{for } x = y \end{cases} \qquad (4.2.19)$$

with boundary values 0. Up to the normalization factor and the minus sign in (4.2.13),
this solution is the Green function as defined in (4.1.31). In particular, the solution
of the Poisson problem

$$\Delta u(x) = g(x) \text{ for } x \in V \backslash V_0 \qquad (4.2.20)$$

$$u(x) = 0 \text{ for } x \in V_0 \qquad (4.2.21)$$

is given by

$$u(x) = - \sum_y v(x, y) g(y), \qquad (4.2.22)$$

that is, the negative of the expected sum of g along the random walk starting at x until it reaches the boundary.

Finally, w from (4.2.2) satisfies

$$w(x, z) = \frac{1}{n_x} \sum_{x' \sim x} w(x', z) \tag{4.2.23}$$

because any path from x to z has to pass through one of the neighbors of x. This means

$$\Delta_x w(x, z) = 0. \tag{4.2.24}$$

For two different boundary points z_1, z_2, we have $w(z_1, z_2) = 0$, and $w(z, z) = 1$. Thus, $w(x, z)$ as a function of x solves the Dirichlet problem (see (4.1.28)) with those boundary values. In words: the probability as a function of the starting point x of the random walk for being absorbed at the boundary point z is the harmonic function u on the graph with boundary values $u(y, z) = \delta(y, z)$. For general boundary values $f(z)$ for $z \in V_0$, the solution of the Dirichlet problem is

$$u(x) = \sum_z w(x, z) f(z). \tag{4.2.25}$$

According to our above interpretation of $w(x, .)$ as the probability distribution for the exit point of the random walk starting at x, we can express (4.2.25) as follows: The solution $u(x)$ at the point x of the Dirichlet problem with boundary values f is the expected value of f at the exit point for the random walk starting at x,

$$u(x) = E(f(w(x, .))). \tag{4.2.26}$$

Obviously, the position X_n of the random walker on our graph constitutes a random process in the sense of Definition 3.2.1. It also satisfies the Markov property of Definition 3.2.2 because the probability distribution for the position X_{n+1} depends only on the location $X_n = x$ at time n, but is independent of earlier positions when given that position at time n.

We now briefly consider the case without boundary, that is, $V_0 = \emptyset$. The transition probabilities for X_n are independent of n and given

$$P(x, y) := p(X_{n+1} = y | X_n = x) = \begin{cases} \dfrac{1}{n_x} & \text{for } y \sim x \\ 0 & \text{for } y \nsim x. \end{cases} \tag{4.2.27}$$

In the above, we have considered the initial distribution $f_0(y) = \delta(x, y)$ (the random walker always started at the point x), but we can obviously consider any initial distribution f_0 with $\sum_y f_0(y) = 1$. Given an initial distribution f_0, the distribution f_n at time n then is $f_0 P^n$ where f_0 is considered as a row vector, that is,

$$f_n(x_n) = \sum_{x_0, x_1, \ldots, x_{n-1}} f_0(x_0) P(x_0, x_1), \ldots, P(x_{n-1}, x_n). \qquad (4.2.28)$$

A distribution π is called stationary if

$$\pi P = \pi, \text{ that is, } \pi(y) = \pi(x) P(x, y). \qquad (4.2.29)$$

The random walk is called ergodic if there exists a unique stationary distribution π with

$$\lim_{n \to \infty} f_0 P^n = \pi \qquad (4.2.30)$$

for every initial distribution f_0. The process is ergodic iff it is irreducible, i.e., for every $x, y \in V$ there exists some n with $P^n(x, y) > 0$, and aperiodic, i.e., the greatest common divisor of the n with $P^n(x, y) > 0$ is 1. The first condition is equivalent to the graph Γ being connected, or in terms of eigenvalues $\lambda_1 > 0$, while the second one is equivalent to Γ being not bipartite, that is, the largest eigenvalue $\lambda_K < 2$, see (2.2.47) and (2.2.48), resp., in 2.2.3.

4.2.2 Diffusion Processes and Partial Differential Equations

We now want to turn to the continuous case (a good reference is [55]). Our heuristic strategy consists in taking a regular lattice as our graph and pass to the continuum limit. This means that we consider the lattice of points $\{h(n_1, \ldots, n_d) : n_1, \ldots, n_d \in \mathbb{Z}\}$ for $h > 0$ which we want to let tend to 0. Thus, our random walker on this lattice when at the lattice point z_n at time n moves to one of its $2d$ lattice neighbors with equal probability $\frac{1}{2d}$. Then the random variable $Z_n = (Z_n^1, \ldots, Z_n^d)$ describing the position of the random walker at time n satisfies

$$Z_n^j - Z_0^j = \sum_{i=1}^{n} X_i \qquad (4.2.31)$$

where the X_i are independent identically distributed random variables with probabilities

$$p(X_i = h) = \frac{1}{2d}, \ p(X_i = -h) = \frac{1}{2d}, \ p(X_i = 0) = \frac{d-1}{d} \qquad (4.2.32)$$

where $X_i = 0$ corresponds to the case where the random step is taking in a direction that is not the ith coordinate direction. The X_i all have expectation value 0 and standard deviation $\frac{1}{d}h^2$. The $Z_n^j - Z_0^j$ then also have expectation value 0, and their standard deviation is $\frac{n}{d}h^2$, by Lemma 3.1.2. By the central limit Theorem 3.1.2, for $n \to \infty$,

$Z_n^j - Z_0^j$ approaches the Gauss distribution $N(0, \frac{1}{d}nh^2) = \frac{1}{\sqrt{2\pi\frac{1}{d}nh^2}}\exp(-\frac{x^2}{2\frac{1}{d}nh^2})$.

We also want to let the size of the time step tend to 0, to compensate for the factor h^2 going to 0. That is, we let the random walker move at times $\tau, 2\tau, 3\tau, \ldots$. Then at time $t = m\tau$, it has jumped m times, and the corresponding position $Z^j(t) - Z^j(0)$ is distributed according to $N(0, \frac{1}{d}\frac{t}{\tau}h^2)$. In order to have a nontrivial limit for positive finite t, we then let h and τ tend to 0 in such a manner that $\frac{h^2}{d\tau} =: \mu^2$ is a positive constant. Thus, in the limit, $Z^j(t) - Z^j(0)$ is distributed according to $N(0, t\mu^2)$. The limiting process $X(t) = (X^1(t), \ldots, X^d(t))$—whose existence one needs to prove—is called the Wiener process or Brownian motion. The components $X^j(t)$ are independent and identically distributed (this is a consequence of the homogeneity of the lattice and the fact that the random walker was moving in each direction with the same probability). $X^j(t + s) - X^j(t)$ is distributed according to $N(0, \mu^2 s)$. In particular, this does not depend on t. Moreover, $X(t_1) - X(s_1)$ and $X(t_2) - X(s_2)$ are independent whenever $s_1 < t_1 < s_2 < t_2$—one says that X has independent increments (cf. the corresponding notion introduced above for point processes). Finally, the typical path $X(t), t \geq 0$ is continuous (but nowhere differentiable).

Again, when we have a bounded domain $\Omega \subset \mathbb{R}^d$ and prescribe continuous boundary values f on $\partial\Omega$, the Dirichlet problem (cf. (4.1.28))

$$\Delta u(x) = 0 \quad \text{for } x \in \Omega \tag{4.2.33}$$

$$u(z) = f(z) \text{ for } z \in \partial\Omega \tag{4.2.34}$$

can be solved by Brownian motion: the (unique) solution $u(x)$ at the point x of the Dirichlet problem with boundary values f is the expected value of f at the exit point for the random walk starting at x,

$$u(x) = E(f(W(x, .))) \tag{4.2.35}$$

where the random variable $W(x, .)$ encodes the exit point from Ω for the random walk starting at x. There is one technical issue here, namely about attaining the boundary values, that is for which points $z \in \partial\Omega$ we have

$$\lim_{x \to z, x \in \Omega} E(f(W(x, z))) = f(z). \tag{4.2.36}$$

The points in $\partial\Omega$ satisfying this condition are called regular. They can be characterized in potential theoretic terms. In particular, this does not depend on the (continuous) function f, but only on the geometry of the domain Ω. Not every point is regular, however. For example, for $d \geq 2$, isolated boundary points are not regular (because they constitute removable singularities for harmonic functions). Here, we do not intend to go into this issue in more detail.

Likewise, up to the minus sign, the Green function is given by the solution $v(x, y)$ of the analogue of (4.2.13) or (4.2.19). In particular, the Poisson problem

$$\Delta u(x) = g(x) \text{ for } x \in \Omega \tag{4.2.37}$$

$$u(z) = 0 \text{ for } z \in \partial\Omega \tag{4.2.38}$$

is given by

$$u(x) = -\int v(x, y)g(y)dy. \tag{4.2.39}$$

Here, $v(x, y)$ is the negative of the Green function, that is, the solution of

$$\Delta_x v(x, y) = -\delta(x, y) \text{ for } x \in \Omega \tag{4.2.40}$$

$$v(x, y) = 0 \text{ for } x \in \partial\Omega, \tag{4.2.41}$$

in analogy to (4.2.19), (4.2.20), (4.2.22).
Similarly, the following interpretation is carried over from the discrete case: For $A \subset \Omega$,

$$v(x, A) := \int_A v(x, y)dy \tag{4.2.42}$$

is the expected amount of time the random walk starting at x spends in A before exiting from Ω. In probabilistic terminology, (4.2.39) is also expressed as

$$u(x) = -E\left(\int_0^{\tau_\Omega} g(X_x(t))dt\right) \tag{4.2.43}$$

where $X_x(t)$ is Brownian motion starting at x and τ_Ω is its expected exit time from Ω. In words: the solution $u(x)$ at x of the Poisson problem for g is given by the negative of the expected integral of g over a random path starting at x until it exits from Ω. In particular, we may put $g = -1$. Then (4.2.43) becomes

$$u(x) = E(\tau_\Omega), \tag{4.2.44}$$

the expected exit time of the random walk starting at x. Thus, this expected exit time is the solution of

$$\Delta u(x) = -1 \text{ in } \Omega, \ u(y) = 0 \text{ for } y \in \partial\Omega. \tag{4.2.45}$$

We return to the probability density

$$P(y, t|x, s) := p(X(t) = y|X(s) = x) = \frac{1}{\sqrt{2\pi(t - s)}} \exp\left(-\frac{(y - x)^2}{2(t - s)}\right) \tag{4.2.46}$$

for $t > s$. This probability density satisfies

$$\frac{\partial P}{\partial t} = \frac{1}{2}\Delta_y P \tag{4.2.47}$$

and

$$\frac{\partial P}{\partial s} = -\frac{1}{2}\Delta_x P \tag{4.2.48}$$

(4.2.47) is called the forward diffusion or Kolmogorov equation, (4.2.48) the backward one. Equation (4.2.47) is also called the Fokker-Planck equation. The interpretation is that the probability density of a stochastic process (here the Wiener process or Brownian motion) satisfies a *deterministic* differential equation.
Equation (4.2.47) is the continuous analogue of (4.2.10). We obtain the Laplace operator here because the lattice that we used for our approximation scheme was regular as all vertices had the same degree $2d$.
We also have an analogue of (4.2.6),

$$P(y, t|x, s) = \int_z P(y, t|z, s + \tau) P(z, s + \tau|x, s)\, dz \text{ for } 0 < \tau < t - s. \tag{4.2.49}$$

Again, the reason for this relation is that whatever arrives at time t at y, originating from x at time s has to be at some point z at the intermediate time $s + \tau$ whence it reaches y after the further time $t - \tau$. (4.2.49) is called the Chapman-Kolmogorov equation. Of course, (4.2.49) can also be derived by a direct computation with Gaussian kernels, on the basis of (4.2.46), but our more abstract derivation is simpler and more insightful. In any case, we again see the ubiquity of Gaussian kernels. By the central limit theorem, our rescaling scheme for the random walk on a lattice produced a Gaussian transition kernel which in turn governs the standard heat equation.
This can also be coupled with a deterministic drift. We consider a general dynamical rule of the form

$$\frac{dy}{dt} = F(y(t), t), \text{ for } y \in \mathbb{R}^d \tag{4.2.50}$$

The continuity equation for the density of y then is

$$\frac{\partial}{\partial t} p(y, t) = -\sum_{i=1}^{d} \frac{\partial}{\partial y^i}(F^i(y, t) p(y, t)). \tag{4.2.51}$$

We now take the sum of Brownian motion and a deterministic dynamics of the form (4.2.50). We write this formally as

$$\frac{dy}{dt} = F(y(t), t) + \eta, \tag{4.2.52}$$

where η is the formal derivative of Brownian motion (which is represented by white noise, but we do not explain this here; see e.g. [69, 87, 96]). (The Eq. (4.2.52) is called the Langevin equation.) By linear superposition of (4.2.47) and (4.2.51), the corresponding density satisfies the Fokker-Planck equation

$$\frac{\partial}{\partial t} p(y, t) = \frac{1}{2} \Delta p(y, t) - \sum_{i=1}^{d} \frac{\partial}{\partial y^i} (F^i(y, t) p(y, t)). \qquad (4.2.53)$$

This issue will be taken up again in 4.5 below.

4.3 Dynamical Systems

4.3.1 Systems of Ordinary Differential Equations

A general reference for this section is [69]. Let $f = (f^1, \ldots, f^n) : \mathbb{R}^n \to \mathbb{R}^n$ be of class C^1. We consider the system of first order ordinary differential equations (ODEs)[2]

$$\dot{x}^i(t) = f^i(x^1(t), \ldots, x^n(t)) \text{ for } i = 1, \ldots, n, \qquad (4.3.1)$$

with $\dot{x}^i = \frac{d}{dt} x^i$. t is considered to be the time, and $x(t) = (x^1(t), \ldots, x^n(t))$ then is the state of the system at time t. One usually prescribes initial values $x_0 = x(0)$ and looks for a solution $x(t), t \in \mathbb{R}(t \geq 0)$. $\{x(t) : t \geq 0\}$ is called the **orbit** of x_0. Equation (4.3.1) is a so-called **autonomous** system because f does not depend explicitly on t (but implicitly through the dependence of x on t).[3] The important point about autonomous systems is that they are invariant under time shifts. This

––––––––––––––––––––

[2] Higher order systems of ODEs can be reduced to systems of first order by introducing additional auxiliary variables.

[3] One may also consider non-autonomous systems,

$$\dot{x}^i(t) = \phi^i(t, x^1(t), \ldots, x^n(t)) \text{ for } i = 1, \ldots, n,$$

with an explicit dependence on t, but such systems can be converted into an autonomous form by introducing a new dependent variable x^{n+1} to obtain the equation

$$\dot{x}^{n+1}(t) = f^{n+1}(x^1(t), \ldots, x^n(t), x^{n+1}(t)) \equiv 1.$$

This may turn linear (non-autonomous) equations into non-linear (autonomous) ones, e.g.

$$\dot{x} = \cos(\beta t)$$

becomes

$$\dot{x}^1 = \cos(\beta x^2)$$
$$\dot{x}^2 = 1$$

means that, if we consider the solution of $(4.3.1)^4$ $x_1(t)$ with initial values $x_1(t_1) = \xi$ and the solution $x_2(t)$ with the same initial values, but starting at time t_2, that is, $x_2(t_2) = \xi$, then for all $t \geq t_2$, $x_2(t) = x_1(t + t_1 - t_2)$. In other words, the behavior of the solution (obviously) depends on the initial values, that is, where or how it starts, but not on the starting time, that is, when it starts.

The Theorem of Picard-Lindelöf yields the short-time existence of solutions:

Theorem 4.3.1 *Suppose that f is Lipschitz continuous, that is, there exists some constant L with*

$$|f(x_1) - f(x_2)| \leq L|x_1 - x_2| \tag{4.3.2}$$

for all $x_1, x_2 \in \mathbb{R}^n$.
For every initial state x_0, the solution $x(t)$ of the system (4.3.1) then exists on some time interval, that is, for

$$-T < t < T, \text{ for some } T > 0.$$

This solution is unique.

The solution need not exist for all times, that is, the maximal such T may be finite. That maximal T in general depends on the initial values x_0. We shall see examples shortly.

An easy, but important consequence of the uniqueness part of the Picard-Lindelöf theorem is that orbits of an autonomous system (4.3.1) cannot intersect or merge. Namely, when at some point x_0 two orbits came together, then there would exist two different solutions (in forward or backward time) with initial values given by x_0.

Since we are imposing no restrictions on f apart from a rather mild smoothness assumption, the behavior of the solutions of systems of ODEs can be rather diverse, and one cannot expect a useful classification. It is more insightful to study certain dynamical motives, that is, qualitative types of behavior of solutions. We start with the case $n = 1$ and write f in place of f^1, and likewise for x. Thus, we look at the scalar equation

$$\dot{x} = f(x) \text{ for } t \geq 0, \text{ with initial condition } x(0) = x_0. \tag{4.3.3}$$

Clearly, there are the simple linear equations, like

$$\dot{x} = 0 \tag{4.3.4}$$

whose solution is constant, $x(t) = x_0$, or

(Footnote 3 continued)
where the dependent variable x^2 enters the r.h.s. non-linearly.
[4] assuming that there exists a unique solution, see below

$$\dot{x} = b \tag{4.3.5}$$

whose solution is linear, $x(t) = x_0 + bt$, or

$$\dot{x} = cx \tag{4.3.6}$$

whose solution is $x(t) = x_0 e^{ct}$. The latter equation, for $c > 0$, already exhibits the important phenomenon that solutions of ODEs can amplify differences over time, that is, when we have two solutions x_1, x_2 with different initial values $x_i(0) = x_{0,i}$, then $|x_1(t) - x_2(t)| = |x_{0,1} - x_{0,2}|e^{ct}$ grows exponentially.

Of course, exponential growth cannot be sustained for a long time. Thus, in many models, one introduces a carrying capacity and considers

$$\dot{x} = cx(m - x) \tag{4.3.7}$$

for $c, m > 0$. This is equation is called the logistic, Verhulst, or Fisher equation. Below, we shall often consider this equation as an example, usually for $c = m = 1$, that is,

$$\dot{x} = x(1 - x). \tag{4.3.8}$$

In (4.3.7), for initial values $0 \le x(0) \le m$, the solution is bounded and stays in that same interval, $0 \le x(t) \le m$ for all $t \ge 0$. In that case, $x(t)$ is monotonically increasing, with $\lim_{t \to \infty} x(t) = m$. When $x(0) > m$, the solution monotonically decays towards the asymptotic value m. $x = m$ and $x = 0$ are both fixed points, that is, when $x_0 = m$ or 0, then $\dot{x}(t) = 0$ for all times, and the solution will stay constant. When the initial values are negative, however, the solution diverges to $-\infty$ in finite time. In particular, we here see the phenomenon that a solution need not exist for all positive times; the simplest example of this is perhaps

$$\dot{x} = x^2, \tag{4.3.9}$$

with the solution

$$x = (\frac{1}{x(0)} - t)^{-1} \tag{4.3.10}$$

which when $x(0)$ is positive becomes infinite in finite time. In fact, the blow-up occurs at $t = \frac{1}{x(0)}$. In contrast, when $x(0) < 0$, the solution exists for all time, with $\lim_{t \to \infty} x(t) = 0$. When $x(0) = 0$, then $x(t) = 0$ for all t. Thus, the fixed point at $x = 0$ separates two different qualitative regimes for the solution of our differential equation.

In general, if $f(x_\star) = 0$ for $i = 1, ..., d$ then $x_\star$ is a fixed point for our Eq. (4.3.3), that is, when $x_0 = x_\star$, then $x(t) = x_\star$ for all t.

Our differential equation (4.3.3) may have several fixed points $x_1 < x_2 < \cdots < x_m$.

(We assume here for simplicity that are only finitely many fixed points. The case of infinitely many fixed points does not lead to substantially new phenomena as the reader will easily check.) If there is no further fixed point between x_k and x_{k+1}, then $f(x)$ cannot have a zero for $x_k < x < x_{k+1}$ and therefore must have a definite sign there. When this sign is positive, then for $x_k < x(0) < x_{k+1}$, the solution $x(t)$ of (4.3.3)—which exists for all t—satisfies $\lim_{t\to\infty} x(t) = x_{k+1}$. Similarly, when $f(x) < 0$ for $x_k < x < x_{k+1}$, the solution with initial values in that interval satisfies $\lim_{t\to\infty} x(t) = x_k$. In particular, the fixed point x_k is attracting when $f(x) > 0$ for $x_{k-1} < x < x_k$ and $f(x) < 0$ for $x_k < x < x_{k+1}$. It is repelling when both signs are reversed. The fixed point 0 for (4.3.9) is neither attracting nor repelling, because $f(x)$ does not change its sign there. When x_m is the largest fixed point, then either $f(x) > 0$ for $x > x_m$ in which case the solution could possibly blow up in finite time, or $f(x) < 0$ for $x > x_m$ in which case the solution monotonically decays to x_m for initial values $x(0) > x_m$. The analogous situation holds when the initial values are smaller than the smallest fixed point.

We can also formulate the following easy global existence result:

Theorem 4.3.2 *We consider* (4.3.3),

$$\dot{x} = f(x) \text{ for } t \geq 0, \text{ with initial condition } x(0) = x_0 \tag{4.3.11}$$

and assume that there exist numbers $m < M$ *with*

$$f(m) > 0 \text{ and } f(M) < 0 \tag{4.3.12}$$

and

$$m \leq x_0 \leq M. \tag{4.3.13}$$

Then (for a Lipschitz continuous f as in (4.3.2)*), the solution of* (4.3.11) *exists for all $t \geq 0$.*

Proof. The key observation is that the solution $x(t)$ has to stay bounded as long as it exists. Whenever it comes near the upper bound M, then $\dot{x} = f(x)$ becomes negative by (4.3.12), and therefore $x(t)$ decreases, and when it comes near the lower bound m, it increases for the same reason. Therefore, we shall have $m \leq x(t) \leq M$ for all t for which the solution exists. By the theorem of Picard-Lindelöf, we can then find some $T > 0$ such that for each t_0 up to which the solution exists, the solution of $\dot{y}(t) = f(y(t))$ with $y(0) = x(t_0)$ exists for $0 \leq t \leq T$. The important point here is that T here does not depend on $x(t_0)$ because the latter is confined in the compact interval $[m, M]$. $x(t) = y(t - t_0)$ then is the solution of (4.3.11) on the interval $[t_0, t_0 + T]$. This implies that the solution has to exist for all time (negative times, although not really our concern, are handled by the same argument). $\qquad\square$

When the r.h.s. of (4.3.3) depends on some parameter λ, that is, we look at

$$\dot{x} = f(x, \lambda) \qquad (4.3.14)$$

then we expect a bifurcation, that is, a qualitative change of behavior of the solutions at those parameter values $\lambda = \lambda_0$ where the number of solutions x_k of

$$f(x, \lambda) = 0 \qquad (4.3.15)$$

changes. For example, for

$$\dot{x} = x^2 + \lambda \qquad (4.3.16)$$

$\lambda = 0$ is such a bifurcation point. For $\lambda > 0$, there is no fixed point, for $\lambda = 0$ there is one, namely 0, and for $\lambda < 0$, there are two, $x = \pm\sqrt{-\lambda}$. Here, we already see the important principle that at generic (that is, typical) bifurcations, fixed always arise or disappear in pairs.

For

$$\dot{x} = \lambda x - x^3 \qquad (4.3.17)$$

$\lambda = 0$ is again a bifurcation point. For $\lambda \leq 0$, $x = 0$ is the only fixed point whereas for $\lambda > 0$, we have additional ones at $x = \pm\sqrt{\lambda}$. The latter ones are attracting whereas 0 is repelling for $\lambda > 0$, but attracting for $\lambda \leq 0$.

The preceding already summarizes the main qualitative results about single ODEs of first order. For $n > 1$, the behavior of solutions of (4.3.1) can become richer and more interesting. When we move to dimension $n = 2$, two new phenomena emerge:

- saddle type fixed points in addition to attracting and repelling ones
- closed periodic orbits.

This is best understood within the context of some wider principles for the analysis of dynamical systems:

1. identify the compact orbits (and perhaps other invariant sets) of the dynamics,
2. linearize about them and
3. investigate their stability.

We now elaborate these points. The simplest case of an invariant set is a fixed point, that is, a point $x_\star$ for which the rhs of (4.3.1) vanishes, that is,

$$f^i(x_\star^1, \ldots, x_\star^n) = 0. \qquad (4.3.18)$$

Typically, the investigation of a dynamical system starts with the identification of these fixed points. We may assume $x_\star = 0$ and study the linearized system

$$\dot{x}(t) = Lx, \quad \text{with } L = \left(\frac{\partial f^i}{\partial x^j}(x_\star) \right)_{i,j=1,\ldots,n}. \qquad (4.3.19)$$

We consider the case $n = 2$ because the above principles already become clear there. The matrix L has either two real eigenvalues or two conjugate complex ones. When it can be diagonalized with two real eigenvalues λ_1 and λ_2, then after a linear change of coordinates, our linearized system becomes

$$\dot{x}^1(t) = \lambda_1 x^1(t)$$
$$\dot{x}^2(t) = \lambda_2 x^2(t), \tag{4.3.20}$$

the solution of which obviously is

$$x^1(t) = e^{\lambda_1 t} x^1(0)$$
$$x^2(t) = e^{\lambda_2 t} x^2(0). \tag{4.3.21}$$

If both eigenvalues are negative, then $x(t)$ converges to the fixed point $x_\star(= 0)$ exponentially while, in the case where both are positive, $x(t)$ exponentially expands.

In the first, attracting, case, $x_\star = 0$ is called a **node** or **sink**, and it is a stable fixed point for $t \to \infty$, whereas in the second, repelling, case, called a **source**, it is unstable for $t \to \infty$. If the two eigenvalues have different signs, say $\lambda_2 < 0 < \lambda_1$, then the fixed point $x_\star = 0$ is neither stable nor unstable. In fact, any initial point on the x^2-axis converges to 0, while all other initial points diverge under the flow. This is called a **saddle**. When one of the eigenvalues vanishes, the picture can get more complicated, and the behaviour of the linearized system may be different from the original one. Actually, this is already seen in the one-dimensional example

$$\dot{x} = x^2 \tag{4.3.22}$$

the linearization of which at 0 is

$$\dot{x} = 0. \tag{4.3.23}$$

When, in contrast to the preceding cases, L has two complex conjugate eigenvalues $\lambda \pm i\mu$, then, after a linear change of coordinates again, we get the system

$$\dot{x}(t) = \begin{pmatrix} \lambda & \mu \\ -\mu & \lambda \end{pmatrix} x(t), \tag{4.3.24}$$

the solution of which is

$$x(t) = e^{\lambda t} \begin{pmatrix} \cos \mu t & \sin \mu t \\ -\sin \mu t & \cos \mu t \end{pmatrix} x(0). \tag{4.3.25}$$

For $\lambda \neq 0$, $x(t)$ moves on a spiral (clockwise or counterclockwise, depending on the sign of μ), exponentially towards 0 for $\lambda < 0$, exponentially expanding for $\lambda > 0$. The case $\lambda = 0$ seems intermediate as the solution then moves on a circle about 0. This case is important for two reasons: it is the first non-trivial example of a compact orbit, and thus an invariant set other than a point. Secondly, this case $\lambda = 0$ is different from all previous cases, insofar as it is not structurally stable. This means that an arbitrarily small variation of $\lambda = 0$ changes the qualitative behavior of the system. Even worse, even the qualitative behavior of the linearized system near the fixed point 0 is no longer the same as that of the original system in case $\lambda = 0$. In a certain sense, these two phenomena are related as we shall try to explain soon. Before doing that, we formulate a general

Definition 4.3.1 A fixed point $x_\star$ of (4.3.1) is called hyperbolic if all eigenvalues of the linearized system have nonvanishing real part.

The qualitative dynamical behaviour near a hyperbolic fixed point is structurally stable in the sense that it is not affected by sufficiently small perturbations or parameter variations (like the eigenvalues of the linearized system), and that it is the same as that of the linearized system—in fact, the difference between the original and the linearized system is such a small perturbation that does not change the qualitative behaviour.
We now look into a non-hyperbolic situation and consider the linear system

$$\dot{x} = y + \lambda x \tag{4.3.26}$$
$$\dot{y} = -x + \lambda y. \tag{4.3.27}$$

The eigenvalues are $\lambda \pm i$, with imaginary part $\neq 0$, and real part $= 0$ for $\lambda = 0$. We may consider λ as a bifurcation parameter, and $\lambda = 0$ as a bifurcation value where the qualitative behaviour of the solution changes. Here, a pair of complex conjugate nonzero eigenvalues crosses the imaginary axis. This is the characteristic criterion for the so-called Hopf bifurcation as we shall now explain.
In the linear system, at $\lambda = 0$ all orbits are periodic, namely circles, about $(0, 0)$, while for $\lambda \neq 0$ there is no periodic orbit at all.
Equation (4.3.26) is the linearization at $(0, 0)$ of

$$\dot{x} = y - x \, (x^2 + y^2 - \lambda) \tag{4.3.28}$$
$$\dot{y} = -x - y \, (x^2 + y^2 - \lambda) \tag{4.3.29}$$

For this system, $(0, 0)$ is a fixed point for all parameter values λ. For $\lambda \neq 0$, the situation is hyperbolic and therefore qualitatively the same as in the linearized system. For $\lambda < 0$, the fixed point is globally exponentially attracting. While this can be deduced from general principles, we can, of course, also see it directly, and this gives us the opportunity to introduce another useful tool, a Lyapunov function which by definition is a function that is strictly decreasing along every flow line. Here, such a Lyapunov function is given by $\log(x^2 + y^2)$ since we have

$$\frac{d}{dt} \log(x^2 + y^2) = 2(-x^2 - y^2 + \lambda) \leq 2\lambda < 0. \tag{4.3.30}$$

Thus, $\log(x^2 + y^2)$ decreases along every flow line, and then so does $x^2 + y^2$, and therefore each flow line has to lead to $(0, 0)$. As already pointed out, this is a structurally stable situation that is invariant under small perturbations of λ.

For $\lambda = 0$, $(0, 0)$ is still globally attracting, but no longer exponentially so. We still have

$$\frac{d}{dt} \log(x^2 + y^2) < 0 \text{ for } (x, y) \neq (0, 0), \tag{4.3.31}$$

but this expression is no longer bounded away from 0. Thus, we see again that the situation at $\lambda = 0$ is not structurally stable.

For $\lambda > 0$, while the situation near $(0, 0)$ is again structurally stable, an interesting global phenomenon emerges away from $(0, 0)$. $(0, 0)$ is repelling, and there exists a periodic orbit $x^2 + y^2 = \lambda$ that is attracting. To understand this, we consider our Lyapunov function:

$$\frac{d}{dt} \log(x^2 + y^2) \begin{cases} > 0 & \text{for } x^2 + y^2 < \lambda \\ = 0 & \text{for } x^2 + y^2 = \lambda \\ < 0 & \text{for } x^2 + y^2 > \lambda. \end{cases} \tag{4.3.32}$$

Thus, when we are on the circle $x^2 + y^2 = \lambda$, we stay there and since, $\dot{x}$ and $\dot{y}$ do not vanish there, it is a nontrivial periodic orbit. When we are outside or inside that circle, we move towards it.

We thus obtain a family, depending on λ, of periodic orbits that emerge from the fixed point at the transition from $\lambda = 0$ to $\lambda > 0$. This family of periodic orbits represents a structurally stable bifurcation, that is, such a family remains under perturbations of the above system.

In contrast to this behaviour, in the linear system, at $\lambda = 0$ all orbits are periodic, namely circles, about $(0, 0)$, while for $\lambda \neq 0$ there is no periodic orbit at all. Thus, here the whole family of periodic orbits is concentrated at a single parameter value, while when the linear system is perturbed by a higher order term, that family gets distributed among different parameter values. The situation at $\lambda = 0$ itself is not structurally stable while the behaviour of the whole family is, namely the emergence of a family of periodic orbits at the transition from an attracting to a repelling fixed point.

We next consider another system with the same linearization (4.3.26), (4.3.27) as the preceding one, (4.3.28), (4.3.29),

$$\dot{x} = y - x \left((x^2 + y^2)^2 - 2(x^2 + y^2) - \lambda \right) \tag{4.3.33}$$
$$\dot{y} = -x - y \left((x^2 + y^2)^2 - 2(x^2 + y^2) - \lambda \right) \tag{4.3.34}$$

depending on a real parameter λ as before. We now have

$$\frac{d}{dt} \log(x^2 + y^2) = 2 \left(-(x^2 + y^2)^2 + 2(x^2 + y^2) + \lambda \right). \tag{4.3.35}$$

This becomes 0 when

$$x^2 + y^2 = 1 \pm \sqrt{1 + \lambda}.$$

Thus, whenever this value is real and nonnegative, we obtain that $x^2 + y^2$ remains constant along a solution, that is, the orbit is a circle. When λ is smaller than -1, no such solution exists. For $\lambda = -1$, we find precisely one solution whereas, for $-1 < \lambda < 0$, we obtain two solutions, of radii $0 < \rho_1 < \rho_2$, say. The right-hand side of (4.3.35) is negative for $0 < \rho := \sqrt{x^2 + y^2} < \rho_1$, but positive for $\rho_1 < \rho < \rho_2$ and negative again beyond ρ_2. Thus, the orbit at ρ_1 is repelling whereas that at ρ_2 is attracting. When λ increases to 0, the repelling periodic orbit at ρ_1 moves into the attracting fixed point at $(0, 0)$. When λ then becomes positive, both the repelling periodic orbit and the attracting fixed point disappear, or, more precisely, the latter turns into a repelling fixed point. Only the attracting periodic orbit at ρ_2 remains. The solution of our system of ODEs then has no option but to move away from the no longer attracting fixed point at $(0, 0)$ to the periodic orbit at ρ_2.

The first bifurcation, that of (4.3.26), (4.3.27), where a stable fixed point continuously changed into a stable periodic orbit was a so-called supercritical Hopf bifurcation. In contrast to this, in a subcritical Hopf bifurcation, as exemplified by (4.3.33), (4.3.34), an unstable periodic orbit coalesces into a stable fixed point so that the latter becomes repelling and no stable orbit is present anymore in its vicinity when the relevant parameter passes the bifurcation value.

The linearization at $(0, 0)$ is the same for both examples, the supercritical and the subcritical Hopf bifurcation. The linearization possesses a pair of complex conjugate eigenvalues whose real parts vanish at the bifurcation point. In fact, by the theorem of E. Hopf, this is precisely the criterion for such a bifurcation where a stable fixed point bifurcates into a family of periodic orbits.

The preceding examples essentially cover the qualitative types of behaviour of two-dimensional systems of ODEs. This is essentially a consequence of the principle observed as a corollary of the Picard-Lindelöf theorem that two orbits can never intersect or merge. In higher dimensions, however, (even though that principle is still in force) the behaviour can get more complicated, and in fact defies a complete classification.

In any case, however, we have an existence result of the type of Theorem 4.3.2.

Theorem 4.3.3 *We consider for $x = (x^1, \ldots, x^n)$ and $f = (f^1, \ldots, f^n)$,*

$$\dot{x} = f(x) \text{ for } t \geq 0, \text{ with initial condition } x(0) = x_0 \tag{4.3.36}$$

and assume that there exist numbers $m^\alpha < M^\alpha$ for $\alpha = 1, \ldots, n$ with

$$f^\alpha(x^1, \ldots, x^{\alpha-1}, m^\alpha, x^{\alpha+1}, \ldots, x^n) > 0 \text{ and} \tag{4.3.37}$$
$$f^\alpha(x^1, \ldots, x^{\alpha-1}, M^\alpha, x^{\alpha+1}, \ldots, x^n) < 0$$

whenever $m^\beta \leq x^\beta \leq M^\beta$ for all $\beta = 1, \ldots, n$, and also

$$m^\beta \leq x_0^\beta \leq M^\beta. \tag{4.3.38}$$

Then (for a Lipschitz continuous f), the solution of (4.3.11) *exists for all $t \geq 0$.*

The *proof* proceeds as that of Theorem 4.3.2.

We now exhibit several systems of ODEs that are important as models at various biological scales.

(1) Biochemical kinetics:
References here are [82, 92]. The basis here is the law of mass action which states that the reaction rate of a chemical reaction is proportional to the concentrations of the reactants raised to the number in which they enter the reaction. That expression is proportional to the collision probability for the reactants. For the simple reaction

$$S_1 + S_2 \rightleftharpoons 2P \tag{4.3.39}$$

when k_+ is the rate constant for the forward reaction that converts $S_1 + S_2$ into $2P$ and k_- is the rate constant for the backward reaction and if we denote the respective concentrations by s_1, s_2, p, then

$$\dot{s}_1 = \dot{s}_2 - = -k_+ s_1 s_2 + k_- p^2 \tag{4.3.40}$$
$$\dot{p} = 2(k_+ s_1 s_2 - k_- p^2). \tag{4.3.41}$$

Enzymatic reactions are of particular importance. The prototype is

$$E + S \rightleftharpoons ES \rightarrow E + P. \tag{4.3.42}$$

Here, the substrate S and the enzyme E first form the enzyme-substrate complex ES in a reversible manner with forward and backward rate constants k_1, k_{-1}, resp., and then the product P is irreversibly released from the enzyme E with rate constant k_2. When we denote the concentrations of E, S, ES, P by e, s, c, p, resp., we obtain the system of ODEs

$$\dot{s} = -k_1 es + k_{-1} c \tag{4.3.43}$$
$$\dot{e} = -k_1 es + (k_{-1} + k_2)c \tag{4.3.44}$$
$$\dot{c} = k_1 es - (k_{-1} + k_2)c \tag{4.3.45}$$
$$\dot{p} = k_2 c. \tag{4.3.46}$$

We observe that p does not appear on the r.h.s of this system. Thus, we need only solve the first 3 equations. p then is obtained by a simple integration. Moreover, the second and third equations are dependent, and we conclude

$$e(t) + c(t) \equiv e_0 \tag{4.3.47}$$

a constant.

Based on the small amount of enzyme needed for such reactions, the Michaelis-Menten theory makes the assumption of a quasi-steady state for the complex ES,

$$\dot{c} = 0. \tag{4.3.48}$$

This is mathematically not unproblematic and requires a singular perturbation analysis, see [92], but here we simply observe the consequence

$$c = \frac{k_1 e_0 s}{k_1 s + k_{-1} + k_2}. \tag{4.3.49}$$

We now move from the molecular to the cellular level and as our next example consider the

(2) Hodgkin-Huxley model for the firing of neurons:
The main variable is the potential V of the neuron, satisfying the ODE

$$C\frac{dV}{dt} = I_e - I_i \tag{4.3.50}$$

where C is the capacitance of the membrane and I_e and I_i are the external and internal currents. The internal current in turn satisfies the equation

$$I_i = g_0(V - V_0) + g_1 m^3 h(V - V_1) + g_2 n^4(V - V_2), \tag{4.3.51}$$

where $g_0, g_1, g_2 > 0$ and V_0, V_1, V_2 are constants whereas m, n, h are gating variables corresponding to the opening of sodium (Na^+) channels leading to the inflow of positively charged Na^+ ions, the opening of potassium (K^+) channels leading to the outflow of positively charged K^+ ions, and the closing of Na^+ channels, resp. Normalizations are such that the gating variables always take their values between 0 and 1 so that the can be interpreted as the probabilities for the corresponding type of channel to be open.
Equations (4.3.50) and (4.3.51) combine to become

$$C\frac{dV}{dt} = I_e - (g_0(V - V_0) + g_1 m^3 h(V - V_1) + g_2 n^4(V - V_2)). \tag{4.3.52}$$

Whereas I_e is treated as an external parameter, the internal dynamical regimes crucially depend on the signs of the three terms in (4.3.52) contributing to I_i. Before going into details, we then formulate a fourth principle for the investigation of systems of ODEs:

4. determine the signs of the diverse summands into which f^i may be decomposed on the right hand side of (4.3.1) and assess their contribution on the global behaviour of the solution.

Before proceeding, however, we need to clarify the roles of the gating variables. m, n, h satisfy differential equations of the form

$$\tau_y(V)\frac{dy}{dt} = y_\infty(V) - y \tag{4.3.53}$$

with the limiting value $y_\infty(V)$ and the time constant $\tau_y(V)$ indicating the time scale on which the corresponding gating variable varies. A simpler model would consist of taking y directly as a function of V, equal to the equilibrium value, that is, $y = y_\infty(V)$. The model of Hodgkin-Huxley instead introduces some additional temporal dynamics where y relaxes to that equilibrium value on the time scale described by $\tau_y(V)$. Thus, in particular, it does not follow changes in V instantaneously, but needs some time to adapt.

The Hodgkin-Huxley systems then consists of 4 differential equations, namely (4.3.52) for the voltage V and three equations of type (4.3.53) for the three gating variables.

It is important for the dynamics of the Hodgkin-Huxley model that while m_∞ and n_∞ are increasing functions of V (m_∞ starts to rise only at a somewhat higher value of V (around $-80\,$mV) than n_∞), h_∞ is a decreasing function. Moreover, the time constant τ_m is much smaller than the time constants τ_n, τ_h (which peak at values of V between 80 and $-70\,$mV), and so m changes much faster than n and h, in fact on the same scale as V.

The reversal potentials in (4.3.51) are

$$V_1 = 50\,\text{mV}$$
$$V_2 = -77\,\text{mV}$$
$$V_0 = -54.4\,\text{mV}.$$

We now present a qualitative discussion of the dynamics of the Hodgkin-Huxley model. Suppose the system initially is at rest near V_0. Then $h = h_\infty(V_0)$ and $n = n_\infty(V_0)$ are positive (in the order of magnitude 1/2) while $m = m_\infty(V_0)$ is close to zero. The relevant term in (4.3.52) then is $g_0(V - V_0)$ which stabilizes the rest point V_0. If now some positive current I_e is injected, a positive feedback dynamics between V and m sets in, as in the range we are entering they are each increasing functions of the other one (recall that m_∞ is an increasing function of V). Namely, once V rises to about $-50\,$mV, m suddenly rises to significantly positive values, and as h is also positive, the Na$^+$ term causes a sharp decrease in the interior current I_i and thus a further rapid increase in V, up to the Na$^+$ equilibrium value of $50\,$mV. Thus, the potential V rises from about -50 to $50\,$mV within a very short time period. This event is called a spike. However, as V rises, h decreases towards 0, and so the Na$^+$ current gets deactivated. In that entire sequence, from the initial rise of m until the decrease of h, the dynamics is essentially driven by the term $g_1 m^3 h(V - V_1)$ in (4.3.52). That term also ensures that the voltage does not exceed the peak value V_1. Moreover, n increases, and so the K$^+$ is activated more strongly, and this causes a decrease of V even below the resting value V_0, down to about V_2, a hyperpolarization. The crucial term for the V dynamics now is $g_2 n^4(V - V_2)$. This causes a refractory period during which no further spike can be fired, during which (in the absence of a further external current) all variables are readjusted back to their resting values.

Let us also describe the above process in physical terms. The dynamics is caused by the interaction of the potential with the inflow of positively charged sodium ions and the outflow of positively charged potassium ions through the activation and inactivation of selective channels in the cell membrane. The sodium channels react more quickly than the potassium ones, so that first by the inflow of positive ions, the cell is depolarized, whereas by the subsequent outflow of positive ions, it gets hyperpolarized. Below threshold, the constant outflow of potassium ions prevents depolarization by a small amount of inflowing sodium ions. When the potential rises above threshold, in the present scenario by some external current, there is a positive feedback between the depolarization (rise of V) and the Na^+ conductance (rise of m), through the voltage triggered opening of sodium ion selective channel in the cell membrane. This causes the rapid spike. However, through the activation of a specific channel protein, the sodium channels close, whereas the potassium channels open, and the cell hyperpolarizes through the outflow of positively charged potassium ions. The details can, for instance, be found in [108]. For more details on the Hodgkin-Huxley model, see [83, 92] and, for new mathematical aspects of it, [100]. The Hodgkin-Huxley model is analyzed with the tools of dynamical systems in much detail in [40, 64], and these are useful references for methods of dynamical systems theory in the neurosciences in general.

It is important to note that already a relatively small or short external current that is barely able to increase V by about 5 mV suffices to trigger the spiking of the neuron, that is, an increase of V by about 100 mV. Thus, a neuron is a device that can amplify the effect of an external input. This input is usually transmitted to a neuron via synaptic connections from other neurons, and one can then study the spreading of activation in a network of neurons.

The Hodgkin-Huxley model is one of the very few biophysical models that not only captures a qualitative behavior, but allows for numerically accurate predictions. It is somewhat complicated, however, in the sense that it is not easy to assess the effects of variations of the parameters involved and that systems of connected Hodgkin-Huxley type neurons become very difficult to analyze. Therefore, at the expense of numerical accuracy, one may seek a simplified model that still captures the important qualitative aspects of spiking neurons. Thus, one seeks a simpler system with the same qualitative behavior of its solutions as the Hodgkin-Huxley model. This starts from the observation that the 4 dependent variables of the Hodgkin-Huxley system evolve on 2 different time scales, a fast one for the evolution of V and m (which both return rapidly to their rest states after a spike) and a slower one for n and h. In particular, since m changes on the same time scale as V itself, it can be taken as a function of the latter and essentially be eliminated from the system. Therefore, one lumps V and m together as a single variable v, and n and $1 - h$ (which show similar behavior) as w. This leads to the FitzHugh-Nagumo system[5] (where we abbreviate $\dot{v} = \frac{dv}{dt}$ etc.)

---

[5] Here, we do not give a detailed derivation of the FitzHugh-Nagumo system from the Hodgkin-Huxley one. See e.g. [83, 92].

$$\dot{v} = v(a - v)(v - 1) - w + \lambda \qquad (4.3.54)$$

$$\dot{w} = bv - cw \qquad (4.3.55)$$

with constants $a \in \mathbb{R}$, $b, c > 0$. The parameter λ here represents the external current I_e, i.e. the input to the neuron.

Thus, the term with m^3 in (4.3.51) translates into the cubic term in (4.3.54). Since the leading coefficient of that cubic term is negative, the dynamics is always confined to some bounded region in the v, w-plane. w enters the system only linearly, and in turn its own evolution equation is linear. The rest point $V = V_0$ in (4.3.51) becomes the origin in (4.3.54), (4.3.55).

It is an open problem to find the explicit solution of this system. Nevertheless, the qualitative aspects of the dynamics can be readily analyzed (see e.g. [92]). We abbreviate

$$f(v) := v(a - v)(v - 1). \qquad (4.3.56)$$

Following the general strategy outlined above for the qualitative analysis of a system of ODEs, we identify the rest points; this is achieved by putting all the time derivatives, that is, the left hand sides of (4.3.54), (4.3.55) equal to 0 and solving the resulting algebraic equation. We start with the analysis for $\lambda = 0$. In that case, the rest points for the FitzHugh-Nagumo system are determined by the equations

$$0 = f(v) - w \qquad (4.3.57)$$

$$0 = bv - cw. \qquad (4.3.58)$$

Depending on the values of the parameters a, b, c, the behavior is described by one of the following two figures (Fig. 4.1)

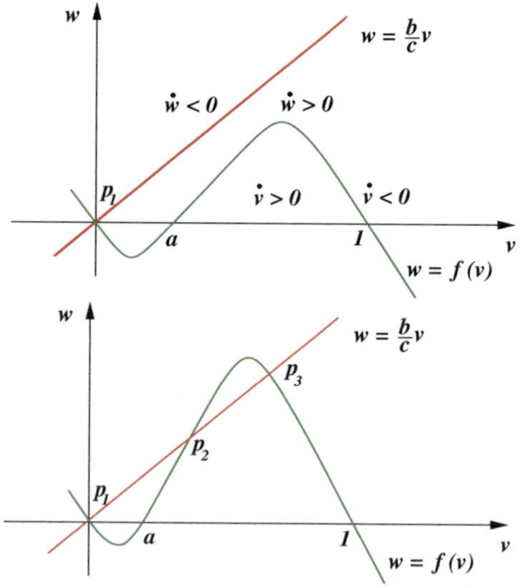

Fig. 4.1 Graphic representation of the FitzHugh-Nagumo equations

Thus, the rest points are at the intersections of the red ($w = \frac{b}{c}v$) and the green ($w = f(v)$) curves. For generic parameter values, we either find one stable fixed point P_1, or two stable ones P_1, P_3 and one unstable one P_2. In the vicinity of the stable fixed point, the behavior of the system is determined by the quadratic term av^2 of $f(v)$ while the negative cubic term $-v^3$ becomes effective only for large values, leading to a resetting of v. Thus, small perturbations of P_1 asymptotically return to P_1. If at $w = 0$, however, v is thrown above the value a, one gets into the region $\dot{v} > 0$, and v thus increases until returning again to the green curve. As one also is in the region $\dot{w} > 0$, w increases as well until reaching the red curve. In the situation captured in the second figure, one then approaches the second stable fixed point P_3. In the situation of the first figure, however, one gets into the region where $\dot{v}$ and $\dot{w}$ both are negative, and v and w thus decrease. The dynamics then gets into the region $v < 0$ left and above the red curve, until $\dot{v}$ eventually becomes positive again, and the dynamics returns to the starting point P_1. This process then is interpreted as the firing of the neuron when the threshold a is exceeded. In summary, we see a qualitatively different behavior, depending on whether the initial perturbation is small or large. In the first case, v directly decreases to its rest point. In the second case, it needs to increase first above a certain value before it is able to return to the rest point. We shall return to the FitzHugh-Nagumo system in Sect. 4.3.2 below where we shall be able understand its dynamics better.

We now wish to analyze the role of the parameter λ that has been left out of the picture so far. After introducing λ, the curve $\dot{v} = 0$ becomes

$$w = f(v) + \lambda; \tag{4.3.59}$$

thus, the green curve is shifted. If λ is positive, from the situation of the first figure, we can either get into that of the second figure (a transition representing a *saddle-node bifurcation* as a stable and an unstable rest point emerge from a contact point between the two curves), or into that of the figure below (Fig. 4.2)

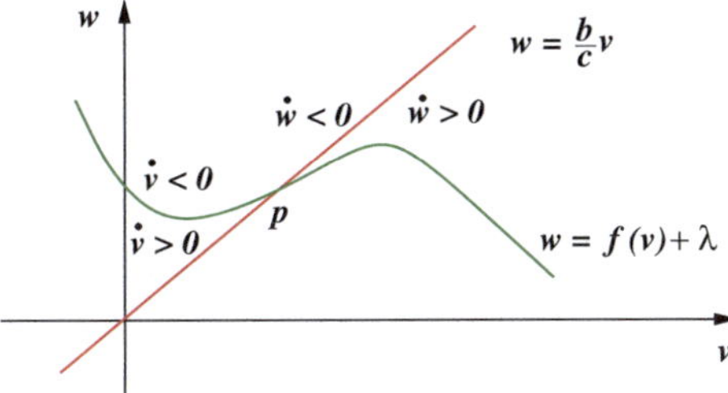

Fig. 4.2 The role of the parameter λ in the FitzHugh-Nagumo equations

In that scenario, the single fixed point P is unstable, and perturbations from the rest position lead first away from P and then turn into oscillations around that rest position as the asymptotic behavior is dominated by the cubic term. We see a Hopf bifurcation, as described above.

Our next example is relevant for a much larger scale, that of ecological interactions of populations:

(3) The Lotka-Volterra system for the sizes x^i of n interacting populations (good references being [60, 61, 92]) is

$$\dot{x}^i = x^i \left(a_i + \sum_{j=1}^{d} b_{ij} x^j\right) \text{ for } i = 1, ..., . \tag{4.3.60}$$

a_i is intrinsic growth or decay rate of the ith population in the absence of the other populations, and b_{ij} is the strength of the effect that the jth population has on the ith one. a_1 is positive (negative) iff x^i has an inherent tendency to grow (decay), and b_{ij} is positive (negative) iff x^j enhances (inhibits) the growth of x^i, e.g. if population i feeds on (is preyed upon by) population j; both b_{ij} and b_{ji} are negative if the two corresponding populations compete. The self-effect b_{ii} is typically negative, expressing a limiting carrying capacity of the environment or interspecific competition for ressources, or at least non-positive. Thus, when x^i gets too large, this term takes over and keeps the population in check.

Biological and other populations always satisfy

$$x^i(t) \geq 0. \tag{4.3.61}$$

Thus, we only need to investigate solutions in the positive quadrant.

For a single population, we consider the logistic or Fisher equation (see (4.3.7))

$$\dot{x}(t) = x(a + bx) \text{ with } a > 0, \ b < 0. \tag{4.3.62}$$

This is about a population growing under the condition of limited or constrained resources, so that, when it gets too large, the capacity limits take over and keep it in balance. $x = 0$ is an unstable fixed point, $x = -a/b$ a stable one.

For the case of two populations, there are three non-trivial scenarios:

1. Predator-prey or parasitism: Population 1 is the prey or host, population 2 its predator or parasite:

   $b_{12} < 0$ (the prey is fed upon by the predators)

   $b_{21} > 0$ (the presence of prey leads to growth of the predator population).

   In the predator-prey scenario, one also typically has

   $a_1 > 0$ (the prey population grows in the absence of predators)

   $a_2 < 0$ (the predator population decays in the absence of prey).

2. Competition

   $b_{12} < 0$ and $b_{21} < 0$ (the two populations inhibit each other).

3. Symbiosis:

$$b_{12} > 0 \text{ and } b_{21} > 0 \text{ (the two populations support each other).}$$

We now look at the example of the two-dimensional predator-prey model without intraspecific competition, that is, we have a prey population of size x^1 and a predator population of size x^2,

$$\dot{x}^1 = x^1(a_1 + b_{12}x^2)$$
$$\dot{x}^2 = x^2(a_2 + b_{21}x^1), \tag{4.3.63}$$

with

$$a_1 > 0, \ a_2 < 0, \ b_{12} < 0, \ b_{21} > 0.$$

We first observe that $(x^1, x^2) = (0, 0)$ is a fixed point. Linearization shows that this fixed point is a saddle. On the x^1-axis, the solution expands according to $x^1(t) = x^1(0)e^{a_1 t}, x^2(t) = 0$, whereas it contracts along the x^2-axis as $a_2 < 0$, $x^1(t) = 0, x^2(t) = x^2(0)e^{a_2 t}$. In particular, since the two axes are orbits, the solution cannot cross them, that is, when starting with non-negative values, it will never turn negative, in accordance with (4.3.61).
Another fixed point is

$$\bar{x}^1 = -\frac{a_2}{b_{21}}, \ \ \bar{x}^2 = -\frac{a_1}{b_{12}} \tag{4.3.64}$$

All the other orbits in the positive quadrant are periodic, circling this fixed point counterclockwise. This is seen either from the local behavior of the trajectories near $(0, 0)$ or by looking at

$$V(x^1, x^2) := b_{21}(\bar{x}^1 \log x^1 - x^1) - b_{12}(\bar{x}^2 \log x^2 - x^2), \tag{4.3.65}$$

which satisfies

$$\frac{d}{dt}V(x^1(t), x^2(t)) = -a_2\frac{\dot{x}^1}{x^1} - b_{21}\dot{x}^1 + a_1\frac{\dot{x}^2}{x^2} + b_{12}\dot{x}^2 \text{ by (4.3.64)}$$
$$= 0 \text{ by (4.3.63).}$$

Thus, $V(x^1, x^2)$ is a constant of motion. V attains its unique maximum at (x^1, x^2), and so the curves $V(x^1, x^2) \equiv$ constant are circles, that is, closed curves, around this point. The motion on such a circle is counterclockwise because in the case $x^1(t) > \bar{x}^1, x^2(t) > \bar{x}^2$ for example, we have $\dot{x}^1(t) < 0, \dot{x}^2(t) > 0$.
On the line $x^1 = \bar{x}^1, \dot{x}^2(t) = 0$, that is, x^2 stays constant there, and on $x^2 = \bar{x}^2, x^1$ stays constant.

Thus, the prey and predator populations oscillate periodically in this model.

The behaviour of the preceding system with its family of periodic orbits is not stable under small perturbations, as we already know from our discussion of the Hopf bifurcation above. For example when we include intraspecific competition, obtaining the system

$$\dot{x}^1 = x^1(a_1 + b_{11}x^1 + b_{12}x^2)$$
$$\dot{x}^2 = x^2(a_2 + b_{21}x^1 + b_{22}x^2) \tag{4.3.66}$$

where we now assume

$b_{11} < 0$ (the members of the prey population compete for food or other ressources)
$b_{22} \leq 0$,

the qualitative behaviour of the system becomes different. We find a second fixed point on the positive x^1-axis, $(-\frac{a_1}{b_{11}}, 0)$. This fixed point is always attractive for x^1, because in case $x^2(t) = 0$, we have the logistic equation (4.3.62)

$$\dot{x}^1(t) = x^1(a_1 + b_{11}x^1) \text{ with } a_1 > 0, \ b_{11} < 0. \tag{4.3.67}$$

It is also attractive for x^2 when

$$a_2 b_{11} - a_1 b_{21} > 0.$$

In that case, there is no other fixed point in the positive quadrant, and in fact for any solution

$$\lim_{t \to \infty} x^2(t) = 0.$$

The predator becomes extinct. Thus, we conclude that a small intraspecific competition among the prey population may lead to the extinction of their predators. If, however,

$$a_2 b_{11} - a_1 b_{21} < 0,$$

then

$$\bar{x}^1 = \frac{a_2 b_{12} - a_1 b_{22}}{b_{11}b_{22} - b_{12}b_{21}} > 0$$
$$\bar{x}^2 = \frac{a_1 b_{21} - a_2 b_{11}}{b_{11}b_{22} - b_{12}b_{21}} > 0$$

is a fixed point in the positive quadrant.
With $V(x^1, x^2)$ as in (4.3.65),

$$\frac{d}{dt}V(x^1(t), x^2(t)) = -b_{11}b_{21}(\bar{x}^1 - x^1(t))^2 + b_{12}b_{22}(\bar{x}^2 - x^2(t))^2 > 0,$$

unless $(x^1, x^2) = (\bar{x}^1, \bar{x}^2)$ in which case this derivative vanishes. Thus, $V(x^1(t), x^2(t))$ increases along every orbit, and equilibrium is possible only at its maximum, at the fixed point $(\bar{x}^1, \bar{x}^2)$. The orbits in the positive quadrant then all spiral counterclockwise towards this fixed point. Thus, in this case, the two populations eventually converge to this fixed point.

We have observed that an arbitrarily small variation of the original system, here by introducing competition among the hosts, changes the global qualitative behavior of the solutions. Therefore, one cannot expect that this model leads to a qualitatively robust and structurally stable behavior, and predictions based on such a model need to be examined with great care. Volterra originally introduced the model to explain the periodic oscillations in two fish populations in the Adriatic, one preying upon the other one. As it turned out, however, this periodic behavior is not caused by an interaction of the two populations according to the model, but rather by periodic changes in the water temperature, that is, by external periodic forcing. Although the model therefore fails its original purpose, it has become useful at a more abstract level, for game theoretic models of interactions inside populations, see [60, 61].

4.3.2 Different Time Scales

For a better understanding of the FitzHugh-Nagumo system (4.3.54) and its capability for modelling the spiking of neurons, it is insightful to look more closely at the fact that the variables v and w change on different time scales—v is fast and w is slow. We therefore now introduce some basic principles for the analysis of dynamical systems with two different time scales. We consider a fast variable x coupled with a slowly changing variable y, satisfying a system of the form

$$\varepsilon \dot{x} = f(x, y, \varepsilon)$$
$$\dot{y} = g(x, y, \varepsilon). \tag{4.3.68}$$

x and y could be scalar or vector valued, f and g are smooth, t is the slow time variable, $\dot{} = \frac{d}{dt}$, and

$$0 < \varepsilon \ll 1. \tag{4.3.69}$$

By transforming to the fast time variable

$$\tau := \frac{t}{\varepsilon}, \tag{4.3.70}$$

with $' = \frac{d}{d\tau}$, we obtain the fast system

$$x' = f(x, y, \varepsilon)$$
$$y' = \varepsilon g(x, y, \varepsilon). \tag{4.3.71}$$

The slow and the fast system differ in the effects resulting from putting $\varepsilon = 0$. From (4.3.68), we obtain the *reduced* problem

$$0 = f(x, y, 0)$$
$$\dot{y} = g(x, y, 0), \tag{4.3.72}$$

whereas (4.3.71) yields the *layered* problem

$$x' = f(x, y, 0)$$
$$y' = 0. \tag{4.3.73}$$

In order to understand the interaction between the slow and the fast time scale, we should therefore analyze the relation and the interplay between the reduced problem (4.3.72) and the layered problem (4.3.73). First, the critical variety S defined by

$$f(x, y, 0) = 0 \tag{4.3.74}$$

yields a constraint for the reduced problem, but represents equilibria for the layered problem. When the derivative $\frac{\partial f(x,y,0)}{\partial x}$ has maximal rank (i.e., is $\neq 0$ if x is a scalar variable), then, by the implicit function theorem (see e.g. [67]), we can locally solve (4.3.74) for $x = h(y)$, that is, transform it into

$$f(h(y), y, 0) = 0. \tag{4.3.75}$$

In this case, S is called normally hyperbolic, otherwise it is non-hyperbolic. In the case where x is scalar, $\frac{\partial f(x,y,0)}{\partial x}$ is also scalar, and its sign decides the stability of S for (4.3.73).

$$\text{If } \frac{\partial f(x, y, 0)}{\partial x} < 0, \quad S \text{ is attracting,}$$
$$\text{if } \frac{\partial f(x, y, 0)}{\partial x} > 0, \quad S \text{ is repelling.} \tag{4.3.76}$$

When x is vector valued, the negative eigenvalues of $\frac{\partial f(x,y,0)}{\partial x}$ correspond to attracting, the positive ones to repelling directions.

The question then is what of this persists for $\varepsilon > 0$. The answer is contained in the Theorem of Fenichel [44].

Theorem 4.3.4 *Assume that S is normally hyperbolic. If $\varepsilon > 0$ is sufficiently small, then there exists a function $h_\varepsilon(y)$ such that the slow manifold*

$$S_\varepsilon = \{(x, y) : x = h_\varepsilon(y)\} \tag{4.3.77}$$

is invariant under (4.3.68) and $O(\varepsilon)$ close to S.

S can loose its normal hyperbolicity due to

1. bifurcation points of S associated with an eigenvalue 0 of $\frac{\partial f(x,y,0)}{\partial x}$ (these may lead to fold points), or
2. different branches of S coming together.

Before applying this scheme to the FitzHugh-Nagumo system (4.3.54), we treat the van der Pol oscillator which is similar, but technically somewhat simpler. This oscillator is described by the system

$$\varepsilon \dot{x} = y - \frac{1}{3}x^3 + x \tag{4.3.78}$$

$$\dot{y} = -x. \tag{4.3.79}$$

First, this system has a fixed point at $(0, 0)$. This fixed point is unstable as the linearization

$$\varepsilon \dot{\xi} = \eta - x^2 \xi + \xi \tag{4.3.80}$$

$$\dot{\eta} = -\xi \tag{4.3.81}$$

at this point has eigenvalues $\frac{1}{2} \pm i\sqrt{\frac{3}{4}}$ which have a positive real part.
The slow manifold S (4.3.74) for (4.3.78), (4.3.79) is given by

$$y = \frac{1}{3}x^3 - x, \tag{4.3.82}$$

and according to (4.3.76), it is attracting for $|x| > 1$, repelling for $|x| < 1$, and not normally hyperbolic for $x = \pm 1$. The slow flow on S is given by

$$(x^2 - 1)\dot{x} = \dot{y} = -x \quad \text{by (4.3.79),} \tag{4.3.83}$$

hence

$$\dot{x} = \frac{x}{1 - x^2}. \tag{4.3.84}$$

This becomes singular at the points $x = \pm 1$ where S looses its normal hyperbolicity. Since the solutions of the system (4.3.78), (4.3.79) cannot escape to infinity, they then exhibit periodic oscillations about the unstable fixed point at $(0, 0)$. For $\varepsilon \ll 1$, they are approximated by the following behavior. For $x > 1$, the dynamics slowly follows the slow manifold S until it reaches the point $x = 1$, $y = -\frac{2}{3}$ whence S becomes unstable, and the dynamics leaves S and quickly moves horizontally to the left to the other point on S with $y = -\frac{2}{3}$ where $x < -1$. There, S is stable again, and

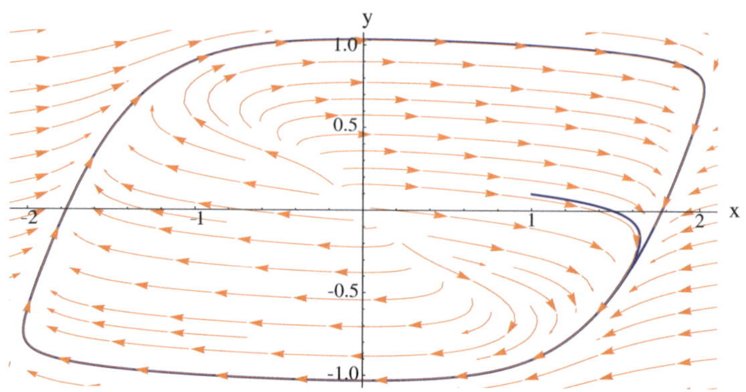

Fig. 4.3 The phase portrait of a van der Pol oscillator for $\varepsilon = .1$

the dynamics therefore slowly follows S again until it reaches $x = -1$, $y = \frac{2}{3}$. Here, S looses again its stability, and the dynamics leaves S and quickly moves horizontally to the right until it hits S again at a value of $y = \frac{2}{3}$ where now $x > 1$. The cycle then starts again. This is displayed in Figs. 4.3 and 4.4.

Another interesting phenomenon occurs if we introduce a parameter λ and consider

$$\varepsilon \dot{x} = y - \frac{1}{3}x^3 + x \qquad\qquad (4.3.85)$$

$$\dot{y} = \lambda - x. \qquad\qquad (4.3.86)$$

This system then has a fixed point at $(x = \lambda, y = \frac{\lambda^3}{3} - \lambda)$. For $\lambda = \pm 1$, this fixed point then is one of the singular points where S is not normally hyperbolic. The linearization (4.3.80), (4.3.81) at this fixed point then has eigenvalues $\frac{1-\lambda^2}{2} \pm \sqrt{\frac{(1-\lambda^2)^2}{4} - 1}$, and their real part vanishes for $\lambda = \pm 1$. At this point, the fixed point then undergoes a Hopf bifurcation. Such a dynamical phenomenon, a Hopf bifurcation at a singular point of the slow manifold, is called a canard explosion, and it leads to an interesting dynamical phenomenon in the limit $\varepsilon \to 0$. Here, however, we do not enter this in more detail.

We shall now apply the slow-fast analysis to the FitzHugh-Nagumo system (4.3.54) (with the control parameter $\lambda = 0$),

$$\varepsilon \dot{v} = v(a - v)(v - 1) - w := f(v, w) \qquad\qquad (4.3.87)$$
$$\dot{w} = bv - cw := g(v, w) \qquad\qquad (4.3.88)$$

with constants $a \in \mathbb{R}$ and $b, c > 0$. That is, v is the fast variable, and w the slow one. Here, we have

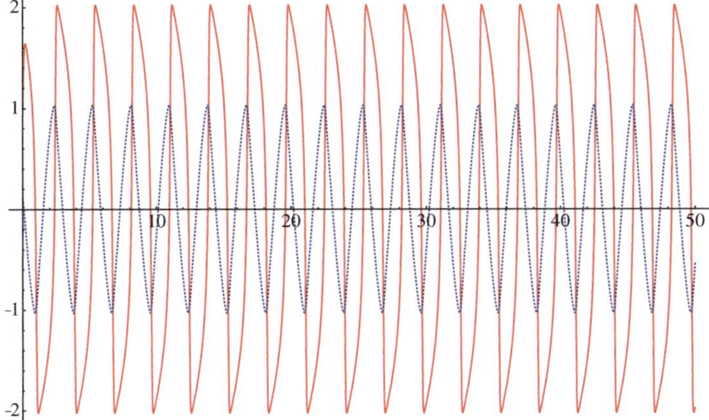

Fig. 4.4 The temporal dynamics of a van der Pol oscillator for $\varepsilon = .1$. *Red $x(t)$, blue $y(t)$*

$$\frac{\partial f}{\partial v} = -3v^2 + 2(a+1)v - a, \tag{4.3.89}$$

and this is

$$> 0 \text{ for } \frac{a+1}{3} - \frac{1}{3}\sqrt{a^2 - a + 1} < v < \frac{a+1}{3} - \frac{1}{3}\sqrt{a^2 - a + 1}$$

$$= 0 \text{ for } v = \frac{a+1}{3} \pm \frac{1}{3}\sqrt{a^2 - a + 1}$$

$$> 0 \text{ for } v < \frac{a+1}{3} - \frac{1}{3}\sqrt{a^2 - a + 1} \text{ or } v > \frac{a+1}{3} - \frac{1}{3}\sqrt{a^2 - a + 1}.$$

Thus, the critical variety S given by

$$v(a - v)(v - 1) - w = 0 \tag{4.3.90}$$

is repelling for $\frac{a+1}{3} - \frac{1}{3}\sqrt{a^2 - a + 1} < v < \frac{a+1}{3} - \frac{1}{3}\sqrt{a^2 - a + 1}$, attracting for $v < \frac{a+1}{3} - \frac{1}{3}\sqrt{a^2 - a + 1}$ and $v > \frac{a+1}{3} - \frac{1}{3}\sqrt{a^2 - a + 1}$, and it is not normally hyperbolic at the two points $v_\pm = \frac{a+1}{3} \pm \frac{1}{3}\sqrt{a^2 - a + 1}$ where the stability properties change. For instance, for $a = 0$, $v_- = 0$, $v_+ = \frac{2}{3}$, whereas for $a = 1$, $v_- = \frac{1}{3}$, $v_+ = 1$. More generally, we have $v_- < 0$ iff $a < 0$. Thus, for $a < 0$, the stable fixed point $v = 0$, $w = 0$ lies between v_- and v_+ on the curve (4.3.90). Therefore, in this case, for small ε, the dynamical picture qualitatively looks as follows. the dynamics detaches from (4.3.90) at v_- before reaching $v = 0$, jumps to the right stable part of (4.3.90), moves along this curve until reaching v_+ where it detaches again to jump to the left stable part of (4.3.90), moves along this curve to v_-, and the cycle repeats itself. Biologically, this is interpreted as the periodic spiking of the neuron modelled by the system (4.3.87), (4.3.88). This behavior is displayed in Figs. 4.5, 4.6 (where

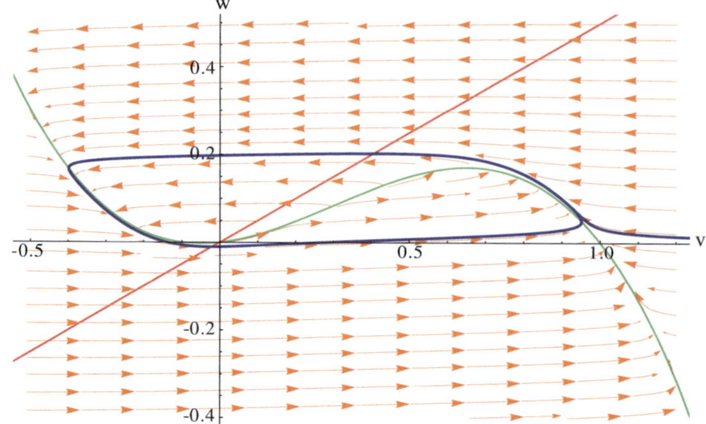

Fig. 4.5 The FitzHugh-Nagumo system for $a = -0.1, b = 0.01, c = 0.02$. *Red line* $w = b/cv$.
Green line $w = f(v)$, *Blue line* graph of $(v(t), w(t))$

Fig. 4.6 A single spike in
the FitzHugh-Nagumo system
for $a = -0.1, b = 0.01, c =
0.02$. *Red line* $v(t)$, *blue line*
$w(t)$

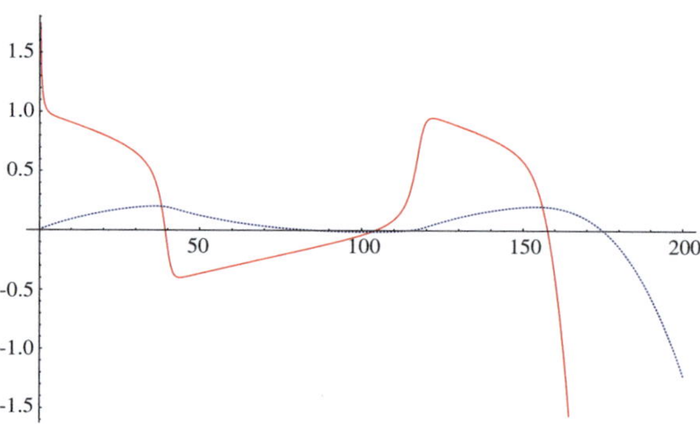

we have put $\varepsilon = 1$, but have chosen b and c very small, which leads to an equivalent
effect as a small ε with b, c of order 1).

In contrast, when $a > 0$, the dynamics may exhibit a single spike, depending on
the initial values, but will eventually move into the stable rest point $v = 0, w = 0$.
This shown in Figs. 4.7, 4.8, 4.9 and 4.10.

For the slow flow on S as given by (4.3.90), we have

$$(-3v^2 + 2(a+1)v - a)\dot{v} = \dot{w} = bv - cw, \tag{4.3.91}$$

whence

$$\dot{v} = \frac{3cv^2 + (b - 2c(a+1))v + ca}{-3v^2 + 2(a+1)v - a}. \tag{4.3.92}$$

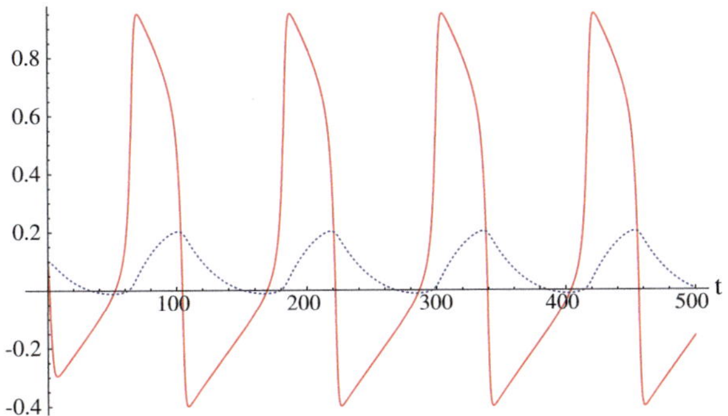

Fig. 4.7 Same dynamics as in Fig. 4.6 for a longer time, showing the periodic spiking pattern

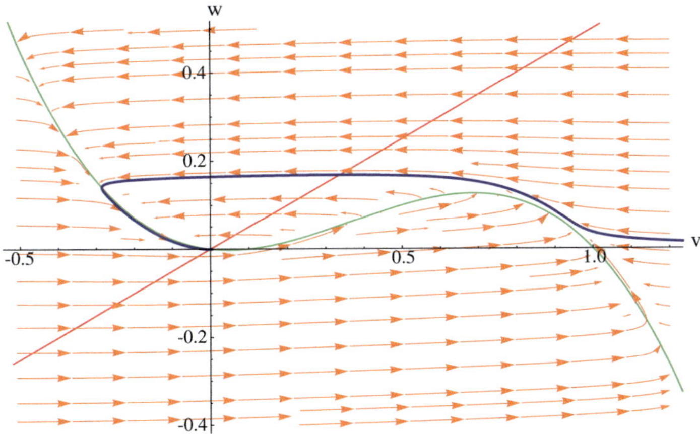

Fig. 4.8 The FitzHugh-Nagumo system for $a = 0.1, b = 0.01, c = 0.02$. *Red line $w = b/cv$.* *Green line $w = f(v)$, Blue line* graph of $(v(t), w(t))$

Fig. 4.9 A single spike in the FitzHugh-Nagumo system for $a = 0.1, b = 0.01, c = 0.02$. *Red line $v(t)$, blue line $w(t)$*

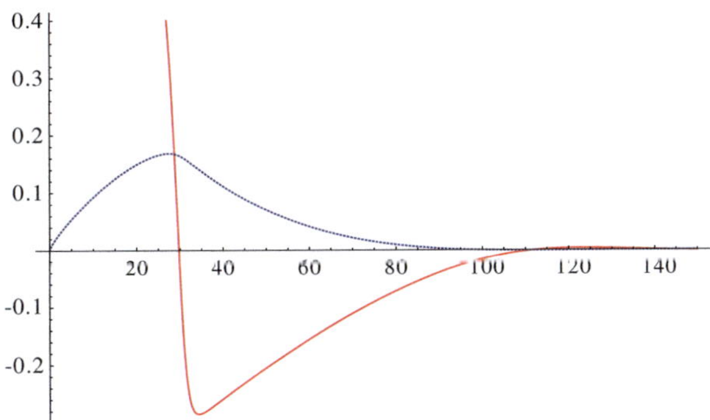

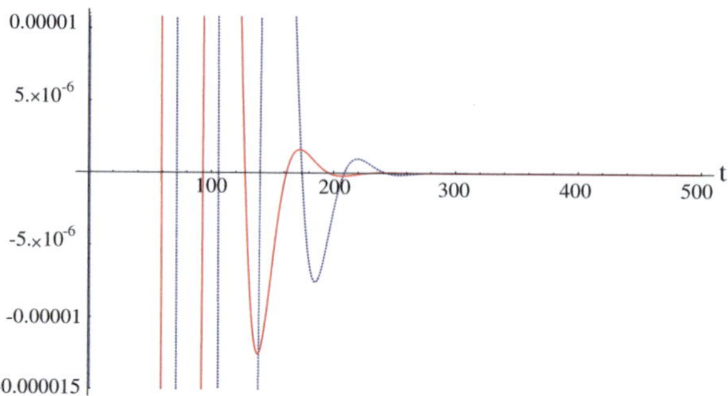

Fig. 4.10 The FitzHugh-Nagumo dynamics with the parameter values of Fig. 4.9 for a longer time, showing the relaxation to the rest point

The slow-fast analysis of dynamical systems with different time scales has found many other applications in biology. Goldbeter has systematically investigated the role of biochemical oscillations that can be captured in ODE models of the type discussed here for the generation of rhythmic cellular behavior, see [52]. For instance, glycolysis is a metabolic process that generates free energy by a chain of enzyme reactions. The intermediates of this process often exhibit oscillatory behavior. The Goldbeter-Lefever model [53] models this via a system of differential equations involving two time scales. A penetrating mathematical analysis of this model along the lines outlined here is developed in [86].

There are many powerful mathematical methods for analyzing the transitions between different scales. While most of these methods have been developed with applications in physics in mind, they could also offer powerful tools for biological problems involving different scales. A good overview of those methods is given in [97].

4.4 Reaction-Diffusion Systems

References for this section include [68, 92, 110].

4.4.1 Reaction-Diffusion Equations

Let $\Omega \subset \mathbb{R}^d$ be open and bounded. We consider the equation

$$u_t(x, t) = \Delta u(x, t) + f(x, t, u) \text{ for } x \in \Omega, 0 < t < T \qquad (4.4.1)$$
$$u(x, 0) = \phi(x) \text{ for } x \in \Omega$$
$$u(y, t) = g(y, t) \text{ for } y \in \partial\Omega, 0 < t < T$$

for continuous and bounded initial and boundary values ϕ, g and a differentiable reaction term f.

We can consider this is a generalization of

1. either the ODE

$$u_t(t) = f(t, u),\tag{4.4.2}$$

(at least in the case where the function f in (4.4.1) does not depend on x), that is, of an equation that does not depend on the spatial variable x and therefore describes a spatially homogeneous state,

2. or the linear heat equation

$$u_t(x, t) = \Delta u(x, t),\tag{4.4.3}$$

that is, of an equation that describes a linear diffusion process in space.

The first equation describes a reaction process, and thus, a reaction-diffusion equation models reaction processes taking place at all points in space simultaneously and being diffusively coupled. It turns out that such an interplay of reaction and diffusion processes can lead to more interesting patterns than either of these processes alone. In order to understand the relationship between the two processes better, we can also introduce a diffusion coefficient d and consider the more general initial value problem

$$u_t(x, t) = d\,\Delta u(x, t) + f(x, t, u), \quad u(x, 0) = \phi(x)\tag{4.4.4}$$

When $d = 0$, we have a system of ODEs indexed by the points x, but without any coupling or interaction between those points. Thus, at each point x, the dynamics is driven by the reaction term, and the result depends only on the initial condition $\phi(x)$ at that particular x. When we let $d \to \infty$, we obtain an equation for the spatially integrated variables, that is, a single ODE for the spatially averaged quantity, and there will be no variation between the different points x. In general, in physical and biological processes, conservation rules will limit growth, and growth at one point x then is only possible at the expense of other points. In principle, a reaction-diffusion equation can then lead to optimal resource allocation in the limit $t \to \infty$. The time scale on which this takes place will depend on the diffusion coefficient d.

Recalling (4.3.8), let us consider

$$u_t = \Delta u + u(1 - u),\tag{4.4.5}$$

for $u = u(x, t)$, that is, the logistic (Verhulst, Fisher) equation with a diffusion term. This equation is sometimes called the Kolmogorov-Fisher equation. It can serve as a model for a population in a uniform habitat with limited capacity that reproduces and diffuses in space. As such, we expect that the behavior of a solution $u(x, t)$ at a point x is not very different from the solution of the ODE $y_t(t) = y(t)(1 - y(t))$

with initial value $y(0) = u(x, 0)$. In particular, we expect that a solution $u(x, t)$ with a positive initial value converges to 1 for $t \to \infty$, unless this prevented by the boundary condition. However, there will be diffusion between the different $x \in \Omega$, and this may decrease the differences in their respective initial conditions faster than the dynamical evolution by the reaction term alone. This effect becomes more important when we consider the spatially inhomogeneous equation

$$u_t(x, t) = \Delta u(x, t) + u(x, t)(a(x) - b(x)u(x, t)) \qquad (4.4.6)$$

for positive functions a, b. Thus, the intrinsic growth rate and the capacity limitations depend on x. Now, the stable equilibrium point for the reaction term, $u = \frac{a(x)}{b(x)}$, depends on the spatial position x. Without diffusion, at every x then in the limit $t \to \infty$, this equilibrium would be obtained. With diffusion, however, we expect some harmonization between the higher and lower values of that equilibrium.

In applications, the dependent variable u typically describes some density, and therefore only non-negative solutions u will be meaningful.

Example:

$$u_t(x, t) = \Delta u(x, t) + u^2(x, t) \text{ for } x \in \Omega, 0 < t \qquad (4.4.7)$$
$$u(x, 0) = \phi(x) \text{ for } x \in \Omega$$
$$u(y, t) = 0 \text{ for } y \in \partial\Omega, 0 < t$$

with

$$\phi > 0 \text{ in } \Omega. \qquad (4.4.8)$$

We recall that the ODE

$$u_t(t) = u^2(t) \qquad (4.4.9)$$

for positive initial value $u(0)$ did blow up in finite time (cf. (4.3.9), (4.3.10)), that is, did not possess a solution that exists for all $t > 0$. We shall now show that the same happens for (4.4.7) provided the initial values ϕ are sufficiently large. To make that condition precise, we recall the first Dirichlet eigenvalue λ_1 and the corresponding eigenfunction u_1 from Theorem 4.1.1, solving

$$\Delta u_1 + \lambda_1 u_1 = 0, \quad u_1 = 0 \text{ on } \partial\Omega,$$

and recall that, by (4.1.68),

$$u_1(x) > 0 \text{ for } x \in \Omega; \qquad (4.4.10)$$

we may normalize u_1 to satisfy

$$\int_\Omega u_1(x)dx = 1. \tag{4.4.11}$$

By the maximum principle Lemma 4.1.2, because of (4.4.8), we have

$$u(x,t) > 0 \text{ for } x \in \Omega, 0 < t. \tag{4.4.12}$$

We look at the auxiliary function

$$y(t) := \int_\Omega u(x,t)u_1(x)dx \tag{4.4.13}$$

which satisfies

$$\begin{aligned}
\dot{y}(t) &= \int_\Omega u_t(x,t)u_1(x)dx = \int_\Omega (\Delta u(x,t) + u^2(x,t))u_1(x)dx \\
&= \int_\Omega (\Delta u_1(x)\, u(x,t) + u^2(x,t)u_1(x))dx = -\lambda_1 y(t) + \int_\Omega u^2(x,t)u_1(x)dx \\
&\geq -\lambda_1 y(t) + y^2(t) \tag{4.4.14}
\end{aligned}$$

since, by the definition (4.4.13), Hölder's inequality,[6] and (4.4.11)

$$y^2(t) \leq (\int u^2(x,t)u_1(x)dx)(\int u_1(x)dx) = \int u^2(x,t)u_1(x)dx.$$

When now

$$y(0) = \int_\Omega u(x,0)u_1(x)dx > \lambda_1, \tag{4.4.15}$$

then (4.4.14) easily implies that $y(t)$ will blow up in finite time, similarly to a solution of (4.4.9) with positive $u(0)$. (More precisely, (4.4.15) implies that $\dot{y}(0) > 0$, and then subsequently $\dot{y}(t) > 0$ for all $t \geq 0$, and the solution will grow and the quadratic term y^2 will dominate the behavior.) This implies that, when (4.4.15) holds, (4.4.7) cannot possess a smooth solution for all positive t.

The maximum principle will allow for a comparison of solutions of a reaction-diffusion equation:

---

[6] Hölder's inequality says that for two L^2 functions f, g (that is, functions with finite $\int f^2(x)dx, \int g^2(x)dx$), we have

$$(\int f(x)g(x)dx)^2 \leq \int f^2(x)dx \int g^2(x)dx.$$

There are also other such calculus inequalities, like the Poincaré and Sobolev ones, which are very useful for the control of solutions of PDEs.

Lemma 4.4.1 *Let u, v be of class C^2 for $x \in \Omega$, $0 < t < T$, and bounded in $\overline{\Omega} \times [0, T]$, and satisfy*

$$u_t - \Delta u - f(x, t, u) \geq v_t - \Delta v - f(x, t, v) \text{ for } x \in \Omega, 0 < t < T \quad (4.4.16)$$

$$u(x, 0) \geq v(x, 0) \text{ for } x \in \Omega, \quad (4.4.17)$$

$$u(y, t) \geq v(y, t) \text{ for } y \in \partial\Omega, 0 < t < T, \quad (4.4.18)$$

or alternatively

$$\frac{\partial u(y, t)}{\partial n} \geq \frac{\partial v(y, t)}{\partial n} \text{ for } y \in \partial\Omega, 0 < t < T. \quad (4.4.19)$$

Then

$$u(x, t) \geq v(x, t) \text{ for } x \in \Omega, 0 \leq t \leq T. \quad (4.4.20)$$

Proof.

$$w(x, t) := u(x, t) - v(x, t)$$

is non-negative for $x \in \partial\Omega, 0 < t < T$ and $x \in \Omega, t - 0$ and satisfies

$$w_t - \Delta w - f_\eta(x, t, \eta)w \geq 0 \quad (4.4.21)$$

for some intermediate $\eta = \epsilon u + (1 - \epsilon)v, 0 \leq \epsilon \leq 1$. The function

$$z(x, t) := w(x, t)e^{-\mu t} \quad (4.4.22)$$

then satisfies

$$e^{\mu t}(z_t - \Delta z - f_\eta(x, t, \eta)z - \mu z) = w_t - \Delta w - f_\eta(x, t, \eta)w \geq 0 \quad (4.4.23)$$

and by making μ sufficiently large, since $f_\eta(x, t, \eta)$ is bounded by the boundedness assumption on u, v, therefore

$$z_t - \Delta z \geq 0 \quad (4.4.24)$$

as long as $z \geq 0$. The strong maximum principle (Lemma 4.1.2) then implies that when z is non-negative and positive somewhere, it cannot become 0 for some $x \in \Omega, t > 0$. Since non-negativity of z implies non-negativity of the difference w of our solutions, this yields the claim. $\qquad\square$

The version of this lemma where an inequality between the exterior normal derivatives of the solutions is assumed allows for a comparison of a solution of (4.4.1) where f does not depend on x, that is,

$$u_t = \Delta u + f(t, u) \quad (4.4.25)$$

$$u(x, 0) = \phi(x) \tag{4.4.26}$$

$$\frac{\partial u(y, t)}{\partial n} = 0 \text{ on } \partial\Omega \tag{4.4.27}$$

with a solution of the corresponding ODE

$$y_t = f(t, y) \tag{4.4.28}$$

$$y(0) = y_0. \tag{4.4.29}$$

When, for example,

$$y_0 \leq \phi(x) \text{ for all } x \in \Omega, \tag{4.4.30}$$

then we conclude that

$$y(t) \leq u(x, t) \text{ for all } x \in \Omega. \tag{4.4.31}$$

Similarly, when y_0 is bigger than ϕ in Ω, then the corresponding solution y of (4.4.28) controls the solution u of (4.4.25) from above.

For example, when

$$f(t, u) = -u^3 \tag{4.4.32}$$

then any solution $y(t)$ of (4.4.28) goes to zero for $t \to \infty$, and when we can then sandwich any solution $u(x, t)$ of (4.4.25) between solutions of (4.4.28) with smaller and larger initial values, resp., than the initial values ϕ of u, and therefore conclude that such a solution u—if it exists for all time—also tends to 0 for $t \to \infty$.

In fact, the general theory of parabolic equations tells us that a solution exists for all time when it can be shown to be bounded. The latter is precisely what can be achieved by such comparison arguments based on the maximum principle. In particular, when solutions of the corresponding reaction ODE stay bounded—and hence exist for all time—for a range of initial values, then so do solutions of the reaction-diffusion equation for the same range of initial values.

Following [27], we shall now derive such a confinement result

Theorem 4.4.1 *We consider a solution u of class* C^2 *of the initial-boundary value problem* (4.4.1)

$$u_t(x, t) = \Delta u(x, t) + f(x, t, u) \text{ for } x \in \Omega, 0 < t < T \tag{4.4.33}$$

$$u(x, 0) = \phi(x) \text{ for } x \in \Omega$$

$$u(y, t) = g(y, t) \text{ for } y \in \partial\Omega, 0 < t < T$$

Suppose that the initial values ϕ *and the boundary values g both satisfy*

$$m \leq \phi(x), g(x, t) \leq M \tag{4.4.34}$$

for all x, t where the numbers m, M satisfy

$$f(x, t, m) > 0 \tag{4.4.35}$$

$$f(x, t, M) < 0. \tag{4.4.36}$$

Then

$$m \leq u(x, t) \leq M \tag{4.4.37}$$

for all $x \in \overline{\Omega}, t \geq 0$. (The solution $u(x, t)$ exists for all time in the present situation when f is assumed to satisfy a Lipschitz bound w.r.t. u; see Sect. 4.4.3 for general statements in this regard.)

Proof.   We shall treat the more general equation

$$u_t(x, t) = \Delta u(x, t) + \sum_{j=1}^{d} h^j(x, t, u)u_{x^j} + f(x, t, u) \tag{4.4.38}$$

that will occur below in Sects. 4.5, 6.2, 6.3, with bounded functions h^j. (For the purpose of the present argument, we do not need to specify regularity assumptions on these h^j. We should keep in mind, however, that our reasoning will require that the solution u be of class C^2, and in order to guarantee that, assumptions like continuity of the h^j are needed.)

We shall consider the upper bound M, the case of the lower one being analogous. Since f is continuous, (4.4.37) continues to hold for $M' = M + \epsilon$ for small ϵ. Therefore, the result follows if we can show that

$$u(x, t) < M' \text{ for all } x \text{ and } 0 \leq t < t_0 \text{ and } u(x_0, t_0) = M' \tag{4.4.39}$$

imply

$$\frac{\partial}{\partial t}u(x_0, t_0) < 0. \tag{4.4.40}$$

Now, that is easy: Since $u(x_0, t_0) \geq u(x, t_0)$ for all x, $u(., t_0)$ achieves a maximum at x_0, and therefore $u_{x^j}(x_0, t_0) = 0$ for all j, as well as $\Delta u(x_0, t_0) \leq 0$. Combined with $f(x, t, M') < 0$, (4.4.38) implies (4.4.40), indeed. $\square$

4.4.2 Travelling Waves

We consider the reaction-diffusion equation in one-dimensional space

$$u_t = u_{xx} + f(u) \tag{4.4.41}$$

and look for solutions of the form

$$u(x, t) = v(x - ct) = v(s), \text{ with } s := x - ct. \tag{4.4.42}$$

This travelling wave solution moves at constant speed c, assumed to be >0 w.l.o.g, in the increasing x-direction. In particular, if we move the coordinate system with speed c, that is, keep $x - ct$ constant, then the solution also stays constant. We do not expect such a solution for every wave speed c, but at most for particular values that then need to be determined.

A travelling wave solution $v(s)$ of (4.4.41) satisfies the ODE

$$v''(s) + cv'(s) + f(v) = 0, \text{ with } ' = \frac{d}{ds}. \tag{4.4.43}$$

When $f \equiv 0$, then a solution must be of the form $v(s) = c_0 + c_1 e^{-cs}$ and therefore becomes unbounded for $s \to -\infty$, that is for $t \to \infty$. In other words, for the heat equation, there is no non-trivial bounded travelling wave. In contrast to this, depending on the precise non-linear structure of f, such travelling waves solutions may exist for reaction-diffusion equations. This is one of the reasons why such equations are interesting.

Example: The Kolmogorov-Fisher equation (4.4.5)

$$u_t = \Delta u + u(1 - u). \tag{4.4.44}$$

It models the spatial spread of a population that grows in an environment with limited carrying capacity. Fisher used it as a model for the spread of an advantageous gene in a population. One then is only interested in non-negative solutions u.

Here, we consider the case where space has only one dimension,

$$u_t = u_{xx} + u(1 - u). \tag{4.4.45}$$

The fixed points of the underlying reaction equation

$$u_t = u(1 - u) \tag{4.4.46}$$

are $u = 0$ and $u = 1$. The first is unstable, the second stable. The travelling wave equation (4.4.43) then is

$$v''(s) + cv'(s) + v(1 - v) = 0. \tag{4.4.47}$$

With $w := v'$, this is converted into the first order system

$$v' = w, \quad w' = -cw - v(1 - v). \tag{4.4.48}$$

The fixed points then are $(0, 0)$ and $(1, 0)$. The eigenvalues of the linearization at $(0, 0)$ (cf. 4.3.1, in particular, the discussion following (4.3.19)) are

$$\lambda_\pm = \frac{1}{2}(-c \pm \sqrt{c^2 - 4}). \tag{4.4.49}$$

For $c^2 \geq 4$, they are both real and negative, and so we obtain a stable node. For $c^2 < 4$, they are conjugate complex with a negative real part, and we obtain a stable spiral. Since a stable spiral oscillates about 0, in that case, we cannot expect a non-negative solution, and so, we do not consider this case here. Also, for symmetry reasons, we may restrict ourselves to the case $c > 0$, and since we want to exclude the spiral then to $c \geq 2$.

The eigenvalues of the linearization at $(1, 0)$ are

$$\lambda_\pm = \frac{1}{2}(-c \pm \sqrt{c^2 + 4}); \tag{4.4.50}$$

they are real and of different signs, and we obtain a saddle. Thus, the stability properties are reversed when compared to (4.4.46) which, of course, results from the fact that $\frac{ds}{dt} = -c$ is negative.

For $c \geq 2$, one finds a solution with $v \geq 0$ from $(1, 0)$ to $(0, 0)$, that is, with $v(-\infty) = 1, v(\infty) = 0$. $v' \leq 0$ for this solution. We recall that the value of a travelling wave solution is constant when $x - ct$ is constant. Thus, in the present case, when time t advances, the values for large negative values of x which are close to 1 are propagated to the whole real line, and for $t \to \infty$, the solution becomes 1 everywhere. In this sense, the behavior of the ODE (4.4.46) where a trajectory goes from the unstable fixed point 0 to the stable fixed point 1 is translated into a travelling wave that spreads a nucleus taking the value 1 for $x = -\infty$ to the entire space.

The question for which initial conditions a solution of (4.4.45) evolves to such a travelling wave, and what the value of c then is, has been widely studied in the literature since the seminal work of Kolmogorov and his coworkers [84]. For example, they showed when $u(x, 0) = 1$ for $x \leq x_1, 0 \leq u(x, 0) \leq 1$ for $x_1 \leq x \leq x_2, u(x, 0) = 0$ for $x \geq x_2$, then the solution $u(x, t)$ evolves towards a travelling wave with speed $c = 2$. In general, the wave speed c depends on the asymptotic behavior of $u(x, 0)$ for $x \to \pm\infty$. Under the assumptions just mentioned, the solution thus converges to the travelling wave with minimal wave speed. To make this more precise, we consider the simplest case

$$u(x, 0) = 1 \text{ for } x < 0, \quad u(x, 0) = 0 \text{ for } x \geq 0. \tag{4.4.51}$$

For each $t > 0$, we then find a unique $\theta(t)$ by

$$u(\theta(t), t) = \frac{1}{2}. \tag{4.4.52}$$

$u(x, t)$ then converges to the travelling wave $v(x - 2t)$ in the sense that

$$u(x + \theta(t), t) \to v(x) \text{ for all } x \in \mathbb{R} \text{ as } t \to \infty. \tag{4.4.53}$$

Since $u_x \theta'(t) + u_t = 0$ and we have the asymptotic relationship $\frac{u_t}{u_x} \to \frac{-2v'}{v'} = -2$ for our wave speed $c = 2$, we also obtain

$$\theta'(t) \to 2 \text{ for } t \to \infty. \tag{4.4.54}$$

This also yields some insights for the asymptotic analysis when we let the diffusion speed go to 0, that is, consider instead of (4.4.45)

$$u_t = \epsilon u_{xx} + u(1 - u). \tag{4.4.55}$$

In this case, assuming again (4.4.51), the minimal wave speed is $c = 2\sqrt{\epsilon}$. The solution u then converges to a travelling wave $v_\epsilon(x - 2\sqrt{\epsilon}t)$ with that wave speed, that is, $u(x + \theta(t), t) \to v_\epsilon(x)$ and $\theta'(t) \to 2\sqrt{\epsilon}$ for $t \to \infty$. Thus, of course, the wave front moves slower and slower as the diffusion speed decreases. Alternatively, one may look at $y(x, \tau) := u(x, \frac{\tau}{\epsilon})$ which then solves

$$y_\tau = y_{xx} + \frac{1}{\epsilon} y(1 - y) \tag{4.4.56}$$

leading to the minimal wave speed $\frac{2}{\sqrt{\epsilon}}$ with which the solution asymptotically moves. Here, we have rescaled the time so that the diffusion speed stays the same, but the reaction term then explodes in the limit, causing faster and faster wave front propagation.

4.4.3 Reaction-Diffusion Systems

We now consider systems of reaction-diffusion equations that are coupled through the reaction terms. These are systems of the form

$$u_t^\alpha(x, t) - d_\alpha \Delta u^\alpha(x, t) = F^\alpha(x, t, u) \quad \text{for } x \in \Omega, t > 0, \alpha = 1, \ldots, n. \tag{4.4.57}$$

Here, $u = (u^1, \ldots, u^n)$ has n components, and the diffusion coefficients d_α are non-negative constants. When some $d_\alpha = 0$, the corresponding equation reduces to an ordinary differential equation for u^α as a function of time t; through the reaction term, it will still be coupled to the other components of u, however, which satisfy partial differential equations when their diffusion coefficients are positive.

For modelling structure formation by systems of reaction-diffusion equations, it is important to allow for different diffusion coefficients d_α (one coefficient $<<1$, another $>>1$): one variable can build spatial concentrations and adapts slowly whereas the other, fast adapting one gets enslaved.

We now state precise existence theorems for the initial-boundary value problem for (4.4.57). The first one is concerned with the existence of solutions for a short time. It is the analogue of the Picard-Lindelöf Theorem 4.3.1. The idea of the proof again is some construction principle. In addition, the maximum principle (see Lemma 4.1.2) plays an important role. This is the reason why now, in contrast to Theorem 4.3.1, we get existence only in forward time.

Theorem 4.4.2 *Let, as always, the diffusion constants d_α all be nonnegative. Let $\Omega \subset \mathbb{R}^d$ be a bounded domain of class C^2, and let*

$$g \in C^0(\partial\Omega \times [0, t_0], \mathbb{R}^n), \quad f \in C^0(\overline{\Omega}, \mathbb{R}^n),$$
$$\text{with } g(x, 0) = f(x) \ \text{ for } x \in \partial\Omega,$$

and let

$$F \in C^0(\overline{\Omega} \times [0, t_0] \times \mathbb{R}^n, \mathbb{R}^n)$$

be Lipschitz continuous w.r.t. u, that is, there exists a constant L with

$$|F(x, t, u_1(x)) - F(x, t, u_2(x))| \leq L|u_1(x) - u_2(x)| \tag{4.4.58}$$

for $x \in \overline{\Omega}, t \in [0, t_0], u_1, u_2 \in \mathbb{R}^n$.
Then there exists some $0 < t_1 \leq t_0$ for which the initial boundary value problem

$$u_t^\alpha(x, t) - d_\alpha \Delta u^\alpha(x, t) = F^\alpha(x, t, u) \ \text{ for } x \in \Omega, \ 0 < t \leq t_1, \alpha = 1, \ldots, n, \tag{4.4.59}$$

$$u(x, t) = g(x, t) \qquad \text{for } x \in \partial\Omega, \ 0 < t \leq t_1,$$
$$u(x, 0) = f(x) \qquad \text{for } x \in \Omega, \tag{4.4.60}$$

admits a unique solution that is continuous on $\overline{\Omega} \times [0, t_1]$.

The proof of this result, while not too difficult and being based on the same principle as that of the Picard-Lindelöf Theorem, that is, a contraction argument for an iteration, is too long to be presented here, and we refer to [68] and the references listed there.

The next result addresses the issue of long-time existence.

Theorem 4.4.3 *We assume that the preceding assumptions hold for all $t_0 < \infty$. We assume furthermore that the solution $u(x, t) = (u^1(x, t), \ldots, u^n(x, t))$ of (4.4.59) satisfies the a-priori bound*

$$\sup_{x \in \overline{\Omega}, 0 \leq \tau \leq t} |u(x, \tau)| \leq K \tag{4.4.61}$$

for all times t for which it exists, with some fixed constant K. Then the solution
$u(x, t)$ *exists for all times* $0 \leq t < \infty$.

This result again follows from the previous, because the boundedness assumption
on the solution makes it possible to apply local existence on time intervals of some
fixed length t_1, independent of u and t, and therefore to keep extending the existence
interval by steps of length t_1.
This naturally leads to the question under which assumptions such an a-priori bound
as in Theorem 4.4.3 holds. This is answered by the analogue of Theorem 4.3.3, that
is, by the extension of Theorem 4.4.1 to systems—the proof of that result extends to
the present case.[7]

Theorem 4.4.4 *Under the above assumptions, suppose that the initial values f and
the boundary values g both satisfy*

$$m_\alpha \leq f^\alpha(x), g^\alpha(x, t) \leq M_\alpha \tag{4.4.62}$$

for all x, t where the numbers m_α, M_α *satisfy*

$$F^\alpha(x, t, u^1, \ldots, u^{\alpha-1}, m_\alpha, u^{\alpha+1}, \ldots, u^n) > 0 \tag{4.4.63}$$

$$F^\alpha(x, t, u^1, \ldots, u^{\alpha-1}, M_\alpha, u^{\alpha+1}, \ldots, u^n) < 0 \tag{4.4.64}$$

whenever $m_\alpha \leq u^\alpha \leq M_\alpha$ *for* $\alpha = 1, \ldots, n$. *Then we have the a-priori bounds*

$$m_\alpha \leq u^\alpha(x, t) \leq M_\alpha \tag{4.4.65}$$

for all $x \in \overline{\Omega}, t \geq 0$. *Consequently, the solution* $u(x, t)$ *exists for all time.*

The region $\{u \in \mathbb{R}^n : m_\alpha \leq u^\alpha \leq M_\alpha \ (\alpha = 1, \ldots, n)\}$ is called an invariant
region for the reaction-diffusion system because a solution that starts in it will never
leave it. The geometric idea behind this is of course that near the lower boundary
value m_α, the component u^α of the solution will increase, because of (4.4.63) and the
maximum principle Lemma 4.1.2, whereas at the upper value M_α, it will decrease.
In other words, the properties of the reaction terms F^α force the solution to stay
inside the region. Therefore, it has to stay bounded.
We now look at an example where Theorem 4.4.4 can be applied: The FitzHugh-
Nagumo equations with diffusion, that is (4.3.54), (4.3.55) for functions $v(x, t)$,
$w(x, t)$ (in place of the notation $u^1(x, t), u^2(x, t)$) that now also depend on a spatial
variable

---

[7] Only the inclusion of derivatives of u as in (4.4.38) would require an additional assumption, to
ensure that $\sum_{\beta=1}^{n} \sum_{j=1}^{d} h^j_{\alpha\beta} u^\beta_{x^j}$ (this is the appropriate generalization of the corresponding term
in (4.4.38)) vanishes whenever u^α assumes a maximum value M or a minimum value m, see [27];
the natural possibility is to assume that each $h^j_{\alpha\beta}$ is a diagonal matrix, that is, the only nonvanishing
entries are of the form $h^j_{\alpha\alpha}$ so that in the αth equation, only derivatives of u^α appear.

$$v_t = \Delta v + v(a - v)(v - 1) - w \tag{4.4.66}$$

$$w_t = \epsilon \Delta w + bv - cw \tag{4.4.67}$$

for some $\epsilon \geq 0$.

We choose m_1, M_1, m_2, M_2 such that (see Fig. 4.1)

1. (m_1, m_2) is below the curves $v(a - v)(v - 1) - w = 0$ and $bv - cw = 0$ (in particular, m_1 and m_2 are both negative)
2. (M_1, m_2) is above the curve $v(a - v)(v - 1) - w = 0$, but below the curve $bv - cw = 0$, and therefore $bv - cw > 0$ whenever $m_1 \leq v \leq M_1$, $w = m_2$
3. (m_1, M_2) is below $v(a - v)(v - 1) - w = 0$, but above the curve $bv - cw = 0$; therefore $v(a - v)(v - 1) - w > 0$ for $v = m_1$, $m_2 \leq w \leq M_2$
4. (M_1, M_2) is above $v(a - v)(v - 1) - w = 0$, and also above $bv - cw = 0$; therefore $v(a - v)(v - 1) - w < 0$ for $v = M_1$, $m_2 \leq w \leq M_2$ as well as $bv - cw < 0$ whenever $m_1 \leq v \leq M_1$, $w = M_2$.

We observe that in fact we can find arbitrarily large rectangles with these properties. Thus, all assumptions of Theorem 4.4.4 are satisfied for arbitrary bounded initial and boundary values, that is, we can always find an invariant region containing them. We conclude the long-time existence of solutions of (4.4.66), (4.4.67) for any such initial and boundary values.

We now turn to the question of when spatial oscillations die out as time tends to infinity, that is, under which conditions the solution of a reaction-diffusion system tends to a spatially homogeneous state. In order to have access to the simplest situation, in place of the Dirichlet boundary conditions that we have used for our existence results, we now assume homogeneous Neumann boundary conditions

$$\frac{\partial u^\alpha(x, t)}{\partial n} = 0 \text{ for } x \in \partial \Omega, \ t > 0, \ \alpha = 1, \ldots, n. \tag{4.4.68}$$

For simplicity, we only discuss the case $F = F(u)$, that is, F is independent of x and t.

Again, we assume that the solution $u(x, t)$ stays bounded and consequently exists for all time. We want to compare $u(x, t)$ with its spatial average $\bar{u}$ defined by

$$\bar{u}^\alpha(t) := \frac{1}{\|\Omega\|} \int_\Omega u^\alpha(x, t) dx \tag{4.4.69}$$

where $\|\Omega\|$ is the volume of Ω.

We also assume a version of (4.4.58):

$$\sup_{x,t} \left\| \frac{dF(u(x, t))}{du} \right\| \leq L. \tag{4.4.70}$$

We let $\lambda_1 > 0$ be the smallest Neumann eigenvalue of Δ on Ω (see Theorem 4.1.2). In order that diffusion can play its role of homogenizing the solution, we need to assume that

$$d := \min_{\alpha=1,\,\ldots,\,n} d_\alpha > 0 \tag{4.4.71}$$

(this d should not be confused with the space dimension).

Theorem 4.4.5 *Under the assumptions just stated, let $u(x, t)$ be a bounded solution of (4.4.57) with homogeneous Neumann boundary conditions (4.4.68). If*

$$\delta := d\lambda_1 - L > 0 \tag{4.4.72}$$

then spatial oscillations of u decay exponentially on average,

$$\int_\Omega \sum_{i=1}^d \sum_{\alpha=1}^n u_{x^i}^\alpha(x, t) u_{x^i}^\alpha(x, t) dx \le c_1 e^{-2\delta t}, \tag{4.4.73}$$

and u approaches its spatial average in the L^2-sense,

$$\int_\Omega |u(x, t) - \bar{u}(t)|^2 dx \le c_2 e^{-2\delta t}. \tag{4.4.74}$$

Here, c_1, c_2 are some constants that depend on the various parameters involved.

Proof. The quantity to consider is

$$E(u(\cdot, t)) = \frac{1}{2} \int_\Omega \sum_{i=1}^d \sum_{\alpha=1}^n u_{x^i}^\alpha u_{x^i}^\alpha dx.$$

(In more condensed notation, $E(u(\cdot, t)) = \frac{1}{2} \int_\Omega |Du(x, t)|^2 dx$.)
We compute its temporal evolution:

$$\frac{d}{dt} E(u(\cdot, t)) = \int_\Omega \sum_{i=1}^d \sum_{\alpha=1}^n u_{tx^i}^\alpha u_{x^i}^\alpha dx$$

$$= \int_\Omega \sum_{i=1}^d \sum_{\alpha=1}^n u_{x^i}^\alpha \frac{\partial(d_\alpha \Delta u^\alpha + F^\alpha(u))}{\partial x^i} dx$$

$$= -\int_\Omega \sum_{\alpha=1}^n d_\alpha \Delta u^\alpha \Delta u^\alpha dx$$

$$+ \int_\Omega \sum_{i=1}^d \sum_{\alpha=1}^n u_{x^i}^\alpha \sum_{\beta=1}^n \frac{\partial F^\alpha}{\partial u^\beta} u_{x^i}^\beta, \text{ since } \frac{\partial u(x, t)}{\partial \nu} = 0 \text{ for } x \in \partial\Omega$$

$$\le (-d\lambda_1 + L) \int_\Omega \sum_{i=1}^d \sum_{\alpha=1}^n u_{x^i}^\alpha u_{x^i}^\alpha dx = -2\delta E(u(\cdot, t)), \tag{4.4.75}$$

using Corollary 4.1.1 and (4.4.72). When we integrate this differential inequality we obtain (4.4.73).

By Corollary 4.1.1, we also have

$$\lambda_1 \int_\Omega |u(x,t) - \bar{u}(t)|^2 dx \leq \int_\Omega \sum_{i=1}^{d} \sum_{\alpha=1}^{n} u_{x^i}^\alpha(x,t) u_{x^i}^\alpha(x,t) dx, \qquad (4.4.76)$$

and therefore, (4.4.73) yields (4.4.74). $\qquad\qquad\square$

We can also derive a similar result for the temporal variation.

Theorem 4.4.6 *Under the same assumptions stated, let $u(x,t)$ again be a bounded solution of (4.4.57) with homogeneous Neumann boundary conditions (4.4.68). If again*

$$\delta = d\lambda_1 - L > 0 \qquad (4.4.77)$$

then temporal oscillations of u decay exponentially on average,

$$\int_\Omega \sum_{\alpha=1}^{n} u_t^\alpha(x,t) u_t^\alpha(x,t) dx \leq c_3 e^{-2\delta t}, \qquad (4.4.78)$$

for some constant c_3.

Proof. The quantity to consider is

$$E_0(u(\cdot,t)) = \frac{1}{2} \int_\Omega \sum_{\alpha=1}^{n} u_t^\alpha u_t^\alpha dx.$$

(In more condensed notation, $E_0(u(\cdot,t)) = \frac{1}{2} \int_\Omega |u_t(x,t)|^2 dx$.)
We compute its temporal evolution:

$$\frac{d}{dt} E_0(u(\cdot,t)) = \int \sum_{\alpha=1}^{n} u_t^\alpha u_{tt}^\alpha$$

$$= \int \sum_{\alpha=1}^{n} u_t^\alpha \frac{\partial}{\partial t}(d_\alpha \Delta u^\alpha + F^\alpha(u))$$

$$= -d_\alpha \int \sum_{i=1}^{d} \sum_{\alpha=1}^{n} u_{x^i t}^\alpha u_{x^i t}^\alpha + \int \sum_{\alpha=1}^{n} u_t^\alpha \sum_{\beta=1}^{n} \frac{\partial F^\alpha}{\partial u^\beta} u_t^\beta$$

$$\leq (-\lambda_1 d + L) \int \sum_{\alpha=1}^{n} u_t^\alpha u_t^\alpha$$

$$= 2(-\lambda_1 d + L) E_0(u(\cdot,t)),$$

using again Corollary 4.1.1, this time for u_t which also satisfies a Neumann boundary condition because u does, and (4.4.77). When we integrate this differential inequality we obtain (4.4.78). $\square$

When all the diffusion constants d_α are equal, one can also establish pointwise decay estimates instead of the coarser L^2-estimates of the preceding theorem.

Theorem 4.4.7 *Let $u(x,t)$ be a bounded solution of*

$$u_t^\alpha(x,t) - \Delta u^\alpha(x,t) = F^\alpha(x,t,u) \ \ for \ x \in \Omega, t > 0 \tag{4.4.79}$$

with homogeneous Neumann boundary conditions

$$\frac{\partial u^\alpha(x,t)}{\partial n} = 0 \ for \ x \in \partial\Omega, \ t > 0, \ \alpha = 1, \ldots, n. \tag{4.4.80}$$

If

$$\delta = \lambda_1 - L > 0 \tag{4.4.81}$$

then u approaches its spatial average exponentially,

$$\sup_{x \in \Omega} |u(x,t) - \bar{u}(t)| \leq c_4 e^{-2\delta t}, \tag{4.4.82}$$

c_4 again being some constant.

For the *proof*, which needs a stronger analytical tools, namely the regularity theory of parabolic partial differential equations and the Sobolev embedding theorem, to convert integral estimates into pointwise ones, we refer to [68]. Similarly, one may obtain a pointwise decay of u_t. Thus, u will tend to a constant as $t \to \infty$, that is, the solution of the reaction-diffusion system will tend towards a homogeneous steady state. Of course, this is not so interesting for pattern formation, and therefore, we now turn to a situation where something else happens.

4.4.4 The Turing Mechanism

The Turing mechanism creates instabilities w. r. t. spatial variables for temporally stable states in a system of two coupled reaction-diffusion equations with different diffusion constants.

The system thus is of the form

$$\begin{aligned}
X_t &= D_X \Delta X + F(X,Y), \\
Y_t &= D_Y \Delta Y + G(X,Y).
\end{aligned} \tag{4.4.83}$$

Here, X is an activator, and Y an inhibitor, and these properties are encoded in the derivatives of F and G, as we shall see below. A key point will be that the inhibitor Y diffuses faster than the activator X, i.e.

$$D_Y > D_X. \tag{4.4.84}$$

Examples:

(1) The Schnakenberg reaction is the mechanism

$$X \rightleftharpoons A, \quad B \to Y, \quad 2X + Y \to 3X \tag{4.4.85}$$

where the concentrations of A and B are kept constant and the concentrations of X and Y are governed by the equations

$$F(X, Y) = k_1 - k_2 X + k_3 X^2 Y,$$
$$G(X, Y) = k_4 - k_3 X^2 Y \tag{4.4.86}$$

where $k_{1,2,3,4} > 0$ are the reaction rates in (4.4.85).

(2) The Gierer-Meinhardt system

$$\begin{aligned} F(X, Y) &= k_1 - k_2 X + \frac{k_3 X^2}{Y}, \\ G(X, Y) &= k_4 X^2 - k_5 Y. \end{aligned} \qquad (k_{1, ..., 5} > 0) \tag{4.4.87}$$

(3) The Thomas system

$$\begin{aligned} F(X, Y) &= k_1 - k_2 X - \frac{k_5 XY}{k_6 + k_7 X + k_8 X^2}, \\ G(X, Y) &= k_3 - k_4 Y - \frac{k_5 XY}{k_6 + k_7 X + k_8 X^2}. \end{aligned} \qquad (k_{1, ..., 8} > 0) \tag{4.4.88}$$

After rescaling the independent and the dependent variables, the system (4.4.83) becomes

$$\begin{aligned} u_t &= \Delta u + \gamma f(u, v), \\ v_t &= d \Delta v + \gamma g(u, v). \end{aligned} \tag{4.4.89}$$

where the parameter $\gamma > 0$ is kept for the subsequent analysis.

The preceding examples then become:

(1)

$$\begin{aligned} u_t &= \Delta u + \gamma(a - u + u^2 v), \\ v_t &= d \Delta v + \gamma(b - u^2 v). \end{aligned}$$

(2)

$$u_t = \Delta u + \gamma\left(a - bu + \frac{u^2}{v}\right),$$
$$v_t = d\Delta v + \gamma(u^2 - v),$$

or slightly more generally

$$u_t = \Delta u + \gamma\left(a - u + \frac{u^2}{v(1 + ku^2)}\right),$$
$$v_t = d\Delta v + \gamma(u^2 - v).$$

(3)

$$u_t = \Delta u + \gamma\left(a - u - \frac{\rho u v}{1 + u + Ku^2}\right),$$
$$v_t = d\Delta v + \gamma\left(\alpha(b - v) - \frac{\rho u v}{1 + u + Ku^2}\right).$$

Here $u, v : \Omega \times \mathbb{R}^+ \to \mathbb{R}$ for some bounded domain $\Omega \subset \mathbb{R}^d$, and we fix the initial values

$$u(x, 0), v(x, 0) \quad \text{for } x \in \Omega,$$

and impose Neumann boundary conditions

$$\frac{\partial u}{\partial n}(x, t) = 0 = \frac{\partial v}{\partial n}(x, t) \quad \text{for all } x \in \partial\Omega, t \geq 0.$$

(We have already seen in Sect. 4.4.1 that Neumann boundary conditions are well adapted for comparing the solutions of reaction-diffusion system with those of the underlying reaction system. One can also study periodic boundary conditions, or, more generally, consider u, v as functions on some compact manifold in place of the domain Ω.)

The mechanism starts with a fixed point $(u_\star, v_\star)$ of the reaction system:

$$f(u_\star, v_\star) = 0 = g(u_\star, v_\star)$$

that is linearly stable. One then investigates its stability under spatially inhomogeneous perturbations. According to the discussion following (4.3.20), the stability as a solution of the reaction system means that all eigenvalues λ of

$$A := \begin{pmatrix} f_u(u_\star, v_\star) & f_v(u_\star, v_\star) \\ g_u(u_\star, v_\star) & g_v(u_\star, v_\star) \end{pmatrix} \tag{4.4.90}$$

have negative real part,

$$\mathrm{Re}(\lambda) < 0. \tag{4.4.91}$$

These eigenvalues are

$$\lambda_{1,2} = \frac{1}{2}\gamma\left((f_u + g_v) \pm \sqrt{(f_u + g_v)^2 - 4(f_u g_v - f_v g_u)}\,\right). \tag{4.4.92}$$

where the derivatives of f and g are evaluated at $(u_\star, v_\star)$. We have $\mathrm{Re}(\lambda_1) < 0$ and $\mathrm{Re}(\lambda_2) < 0$ if

$$f_u + g_v < 0, \tag{4.4.93}$$
$$f_u g_v - f_v g_u > 0. \tag{4.4.94}$$

Since the diffusion terms are already linear, we thus obtain the linearization of the reaction-diffusion system about $(u_\star, v_\star)$ as

$$w_t = \begin{pmatrix} 1 & 0 \\ 0 & d \end{pmatrix} \Delta w + \gamma A w. \tag{4.4.95}$$

According to Theorem 4.1.2, we let u_k be an orthornormal basis of eigenfunctions of Δ on Ω with Neumann boundary conditions and corresponding eigenvalues λ_k,

$$\Delta u_k + \lambda_k u_k = 0 \quad \text{in } \Omega,$$
$$\frac{\partial u_k}{\partial n} = 0 \quad \text{on } \partial\Omega. \tag{4.4.96}$$

We then look for solutions of (4.4.95) of the form

$$w_k e^{\lambda t} = \begin{pmatrix} \alpha u_k \\ \beta u_k \end{pmatrix} e^{\lambda t}. \tag{4.4.97}$$

Inserting this into (4.4.95) yields

$$\lambda w_k = -\begin{pmatrix} 1 & 0 \\ 0 & d \end{pmatrix} \lambda_k w_k + \gamma A w_k. \tag{4.4.98}$$

This means that λ thus should be an eigenvalue of

$$\left(\gamma A - \begin{pmatrix} 1 & 0 \\ 0 & d \end{pmatrix} \lambda_k\right).$$

Written out, the eigenvalue equation is

$$\lambda^2 + \lambda(\lambda_k(1+d) - \gamma(f_u + g_v))$$
$$+ d\lambda_k{}^2 - \gamma(df_u + g_v)\lambda_k + \gamma^2(f_u g_v - f_v g_u) = 0. \tag{4.4.99}$$

As a quadratic equation, this has two solutions $\lambda(k)_{1,2}$. Since $\lambda_0 = 0$, (4.4.93), (4.4.94) imply

$$\mathrm{Re}\ \lambda(0)_{1,2} < 0.$$

We now wish to investigate whether we can have

$$\mathrm{Re}\ \lambda(k) > 0 \tag{4.4.100}$$

for some higher mode λ_k.

Applying (4.4.93) again, since $\lambda_k > 0$, $d > 0$, we have

$$\lambda_k(1+d) - \gamma(f_u + g_v) > 0.$$

Thus for (4.4.100), we should have

$$d\lambda_k{}^2 - \gamma(df_u + g_v)\lambda_k + \gamma^2(f_u g_v - f_v g_u) < 0. \tag{4.4.101}$$

Using now (4.4.94), this can only happen if

$$df_u + g_v > 0.$$

If we assume

$$f_u > 0, \quad g_v < 0, \tag{4.4.102}$$

then this is only compatible with (4.4.93) if

$$d > 1. \tag{4.4.103}$$

We are now ready to show

Lemma 4.4.2 *Let* (4.4.102) *and* (4.4.103) *hold and suppose*

$$\frac{(df_u + g_v)^2}{4d} > f_u g_v - f_v g_u. \tag{4.4.104}$$

Then there exist $\mu_- < \mu_+$ such that $(u_\star, v_\star)$ is spatially unstable w. r. t. the mode λ_k whenever $\mu_- < \lambda_k < \mu_+$.

Proof. We need to find a solution of (4.4.98) with

$$\mathrm{Re}\ \lambda > 0.$$

We have reduced this condition to (4.4.101). Thus, we need to find a nontrivial range of λ_k for which this holds.

The value μ of λ_k for which the left hand side of (4.4.101) becomes minimal is

$$\mu = \frac{\gamma}{2d}(df_u + g_v). \tag{4.4.105}$$

When (4.4.104) holds, the left hand side of (4.4.101) is negative for $\lambda_k = \mu$ and vanishes for

$$\mu_\pm = \frac{\gamma}{2d}\left((df_u + g_v) \pm \sqrt{(df_u + g_v)^2 - 4d(f_u g_v - f_v g_u)}\right) \tag{4.4.106}$$

$$= \frac{\gamma}{2d}\left((df_u + g_v) \pm \sqrt{(df_u - g_v)^2 + 4df_v g_u}\right)$$

Therefore, it is negative in the range

$$\mu_- < \lambda_k < \mu_+. \tag{4.4.107}$$

$\square$

The question now is when (4.4.104) is satisfied and when one or several of the eigenvalues of Δ are contained in the range given by (4.4.107). For the first issue, we need

$$d > d_c = -\frac{2f_v g_u - f_u g_v}{f_u^2} + \frac{2}{f_u^2}\sqrt{f_v g_u(f_v g_u - f_u g_v)}\,. \tag{4.4.108}$$

For the second issue, i.e., whether we can find an eigenvalue λ_k in the range (4.4.107), the geometry of Ω comes into play. In order to understand this better, we recall the discussion after Theorem 4.1.2. This tells us that the eigenvalues scale like $\|\Omega\|^{-\frac{2}{d}}$. Therefore, for a small Ω, all nonzero eigenvalues are large. In particular, we can make all nonzero eigenvalues are larger than μ_+ if we take Ω sufficiently small. Alternatively, we may keep Ω, and hence its smallest nonzero eigenvalue λ_1, fixed and choose $\gamma > 0$ in (4.4.106) so small that

$$\mu_+ < \lambda_1.$$

Then, again, (4.4.107) cannot be solved. From these considerations we see that we need a certain minimal domain size for a given reaction strength, or else a certain minimal reaction strength for a given domain size, for a Turing instability to occur.

Let us now assume that (4.4.107) is satisfied for some eigenvalue λ_k. By Courant's nodal domain theorem, the nodal set $\{u_k = 0\}$ of the eigenfunction u_k divides Ω into at most $(k + 1)$ regions. On any of these regions, u_k then has a fixed sign, i.e. is either positive or negative on that entire region. Since u_k is the unstable mode, the value of k therefore controls the number of oscillations of the developing instability.

We summarize our considerations in

Theorem 4.4.8 *Suppose that at a solution $(u_\star, v_\star)$ of*

$$f(u_\star, v_\star) = 0 = g(u_\star, v_\star),$$

we have

$$f_u + g_v < 0, \quad f_u g_v - f_v g_u > 0. \tag{4.4.109}$$

Furthermore, suppose that $d > 1$ satisfies

$$d f_u + g_v > 0, \quad (d f_u + g_v)^2 - 4d(f_u g_v - f_v g_u) > 0. \tag{4.4.110}$$

By (4.4.109), on one hand $(u_\star, v_\star)$ is linearly stable for the reaction system

$$u_t = \gamma f(u, v),$$
$$v_t = \gamma g(u, v),$$

but by (4.4.110), on the other hand, it is linearly unstable as a solution of the reaction-diffusion system

$$u_t = \Delta u + \gamma f(u, v),$$
$$v_t = d \Delta v + \gamma g(u, v)$$

against spatial oscillations with eigenvalue λ_k whenever λ_k is contained in the range given by (4.4.107).

Thus, in the situation described in Theorem 4.4.8, the equilibrium state $(u_\star, v_\star)$ is unstable, and in the vicinity of it, perturbations grow at a rate $e^{\mathrm{Re}\lambda}$, where λ solves (4.4.99).

As we have already discussed, since the bounded domain Ω has a discrete sequence of eigenvalues λ_k of Δ by Theorem 4.1.2, it depends on the geometry of Ω whether we can find an eigenvalue within the range identified by (4.4.107). If so, the number k controls the frequency of oscillations of the instability about $(u_\star, v_\star)$, and thus determines the shape of the resulting spatial pattern.

In any biological application, it is natural to require that the dynamics stays bounded, and typically, it should also be nonnegative. For that purpose, we need assumptions on f and g for $u = 0$ or $v = 0$, or for u and v large that should ensure that a solution that starts in the positive quadrant can neither become zero nor unbounded. In Sect. 4.4.3 we have discussed the principle that if such a confinement holds for the reaction system, it also holds for the reaction-diffusion system. Thus, even though $(u_\star, v_\star)$ is locally unstable, and therefore small perturbations grow exponentially, this growth has to stop eventually. In fact, in many cases of biological interest, the corresponding solution of the reaction-diffusion system should settle at some spatially inhomogeneous steady state. This latter point has not yet not been settled conclusively. As far as the author knows, the existence of a spatially heterogeneous solution has only been shown by singular perturbation analysis near the critical parameter d_c in (4.4.108).

We should also point out that Turing structures show that what we have discussed at the end of Sect. 4.4.3, namely that solutions of reaction-diffusion systems become spatially homogeneous as time tends to infinity, is by no means a universal phenomenon, but rather depends on specific assumptions. Clearly, spatially inhomogeneous structures as produced by the Turing mechanism are more interesting for pattern formation than homogeneous ones. Of course, the situation becomes even richer when the asymptotic structures are not only not spatially homogeneous, but also not steady in time. For example, a Turing instability could get combined with a Hopf bifurcation. See [120] for examples.

Equipped with Theorem 4.4.8, we return to the example (1), the Schnakenberg reaction. We find $a, b > 0$ with

$$u_\star = a + b,$$

$$v_\star = \frac{b}{(a+b)^2}.$$

At these values $(u_\star, v_\star)$

$$f_u = \frac{b-a}{a+b},$$

$$f_v = (a+b)^2,$$

$$g_u = -\frac{2b}{a+b},$$

$$g_v = -(a+b)^2,$$

$$f_u g_v - f_v g_u = (a+b)^2 > 0.$$

We can therefore translate the conditions of Theorem 4.4.8 into inequalities between a, b and d. Since for the condition $df_u + g_v > 0$, f_u and g_v must have opposite signs, we stipulate

$$b > a.$$

$f_u + g_v < 0$ then requires

$$0 < b - a < (a+b)^3, \tag{4.4.111}$$

and $df_u + g_v > 0$ needs

$$d(b-a) > (a+b)^3. \tag{4.4.112}$$

In order to satisfy $(df_u + g_v)^2 - 4d(f_u g_v - f_v g_u) > 0$, we finally need

$$\left(d(b - a) - (a + b)^3\right)^2 > 4d(a + b)^4. \tag{4.4.113}$$

These inequalities can be easily achieved, essentially by choosing d sufficiently large. And when they hold, the conditions for the Turing phenomenon are met, and we only need a domain Ω with an eigenvalue of the Laplacian in the right range as described in Theorem 4.4.8.

The Turing mechanism is a beautiful analytical scheme for pattern formation. This, however, does not imply that this really is the general scheme underlying the formation of spatial patterns in biology. In fact, according to recent developments in developmental biology, the combinatorial patterns of gene regulation constitute the basic mechanism for the formation of spatial structures, rather than the Turing mechanism. Nevertheless, in certain cases, Turing's idea [117] may apply. Many examples are discussed in [92]. A Turing type mechanism may play an important role in cell division by organizing the Min proteins that determine the localization of the division site [89].

The present section follows the presentation in [68] rather closely.

4.5 Diffusion and Continuity Equations

We assume that we have a variable u describing the state of a biological (or other) system that takes values in $\mathbb{R}^n$ (or some other space) and that changes as a function of time t and perhaps also of some spatial variable x, say $x \in \Omega \subset \mathbb{R}^d$. The two different types of PDE models in biology then correspond to whether we want to investigate

1. the state as a function of space and time, that is, derive an equation for $u(x, t)$, or
2. the distribution of states as a function of time, that is, consider some density or other function $h(u, t)$, so that u now is an independent variable; for instance, in the spatial case, $h(u, t)$ could be the density function for the points x that are in state u at time t.

In either case, the model can start with a dynamical state equation of the form

$$u_t(t) = f(t, u) \tag{4.5.1}$$

and then add some stochastic effects leading to a diffusion. In the first case, this leads to diffusion in physical space, that is, the state value $u(x, t)$ is diffusing to neighboring points, and we obtain a reaction-diffusion system as in (4.4.57)

$$u_t^\alpha(x, t) - d_\alpha \Delta u^\alpha(x, t) = f^\alpha(x, t, u) \quad \text{for } x \in \Omega, t > 0, \alpha = 1, \ldots, n. \tag{4.5.2}$$

Here, all diffusion coefficients d_α are assumed to be nonnegative. When we assume that all the $d_\alpha = 1$, and if we suppress the index α, we can write the system as the prototype

$$u_t(x, t) - \Delta u(x, t) = f(x, t, u). \tag{4.5.3}$$

As explained in Sects. 4.4.1, 4.4.3, the main tool for studying such systems is the maximum principle.

Of course, one can allow for more general diffusion schemes, leading to different diffusion constants d_α as in the Turing model, or even to diffusion operators other than the Laplacian Δ.

In the second case, the object of interest then is the density $h(u, t)$ of u. That density $h(u, t)$ could be a probability density, that is, for each measurable $A \subset \mathbb{R}^n$, the probability that $u(t)$ is contained in A is given by

$$\int_A h(y, t)dy, \tag{4.5.4}$$

and we have the normalization

$$\int_{\mathbb{R}^n} h(y, t)dy = 1 \text{ for all } t \geq 0. \tag{4.5.5}$$

In order to satisfy that normalization (4.5.5), we may assume that it satisfies that normalization initially,

$$\int_{\mathbb{R}^n} h(y, 0)dy = 1, \tag{4.5.6}$$

and then let it evolve according to the continuity equation

$$\frac{\partial}{\partial t}h(u, t) = \sum_{i=1}^{n} \frac{\partial}{\partial u^i}(-f^i(t, u)h(u, t)) \tag{4.5.7}$$

(together with some suitable decay at infinity, for technical reasons).

This equation states that the change of the probability density in time is the negative of the change of the state as a function of its value.

Equation (4.5.7) is a first order partial differential equation of hyperbolic type. When $f = 1$, we obtain a so-called transport equation

$$\frac{\partial}{\partial t}h(u, t) + \frac{\partial}{\partial u}h(u, t) = 0 \tag{4.5.8}$$

(suppressing the index i for simplicity).

As a simple biological example, we consider the case $n = 1$ and let u stand for the age of individuals in a population. For purposes of idealization, we assume that the population is infinite, so that we can have densities, and there are no birth and death processes. $h(u, t)$ then is the fraction of the population of age u at time t. Equation (4.5.8) then describes the change of that density as individuals get older. When we start with the dynamical system, the solution of (4.5.7) depends only on the initial values $u(0)$, and as such, we expect $h(u, t)$ to evolve as $\delta(u - u(t))$, where δ is the Dirac functional and $u(t)$ is the solution of (4.5.1). In the dynamical systems setting, this formalism then becomes more interesting when we consider the simultaneous evolution of a family of initial values, instead of only a single value $u(0)$. If we are in the spatial situation, of course, $h(u, t)$ can simply be the fraction of points x that are in state u at time t, as already explained. We can consider, however, (4.5.7) also in situations without spatial distributions, and rather consider the dynamical evolution of a distribution of initial states. This then represents an important paradigm shift for dynamical systems. Instead of focussing on individual trajectories, that is, considering the evolution of a single initial value in time, we now rather take a family of initial values and investigate how the density of states evolves. The limiting density, if it exists, then represents a stationary state distribution.

In order to treat (4.5.7) mathematically, we describe the method of characteristics. For that purpose, we first consider a somewhat different system,

$$\frac{\partial}{\partial t} h(u, t) + \sum_{i=1}^{n} f^i(t, u) \frac{\partial h(u, t)}{\partial u^i} = 0 \tag{4.5.9}$$

with prescribed initial values $h(u, 0)$.
We then consider the characteristic equation

$$\begin{aligned} Y_t(t, u) &= f(t, Y(t, u)) \\ Y(0, u) &= u. \end{aligned} \tag{4.5.10}$$

We then have

Lemma 4.5.1 *Suppose that f is of class C^1, and satisfies the bounds*

$$\begin{aligned} |f(t, u)| &\leq c_1(1 + |u|) \tag{4.5.11} \\ |f(t, u) - f(t, v)| &\leq c_2|u - v| \text{ for all } t \in \mathbb{R}, u, v \in \mathbb{R}^n. \tag{4.5.12} \end{aligned}$$

Then for initial values $h(u, 0)$ of class C^1, there exists a unique solution $h(u, t)$ of (4.5.9) of class C^1. This solution is characterized by the property that it is constant along characteristics, that is,

$$h(Y(t, u), t) = h(u, 0) \text{ for all } t \in \mathbb{R}, u \in \mathbb{R}^n. \tag{4.5.13}$$

This solution then satisfies

$$\inf_{v \in \mathbb{R}^n} h(v, 0) \leq h(u, t) \leq \sup_{v \in \mathbb{R}^n} h(v, 0) \text{ for all } t, u. \tag{4.5.14}$$

Proof.

$$\frac{d}{dt}h(Y(t,u),t) = \frac{\partial}{\partial t}h(Y(t,u),t) + \sum_{i=1}^{n} Y_t^i(t,u)\frac{\partial}{\partial u^i}h(Y(t,u),t)$$

$$= \frac{\partial}{\partial t}h(Y(t,u),t) + \sum_{i=1}^{n} f^i(t,Y(t,u))\frac{\partial}{\partial u^i}h(Y(t,u),t).$$

Thus, (4.5.9) is satisfied if $h(Y(t,u),t)$ is independent of t, and the initial condition then yields (4.5.13). The assumptions on f guarantee the existence of the solution Y of the characteristic equation. The bound (4.5.14) is obvious. $\square$

The solution of (4.5.9) given by (4.5.13) is no more regular than its initial values $h(u,0)$. Subsequently, when we introduce diffusion, we shall obtain improved regularity, but before doing that, we return to (4.5.7). We rewrite that equation as

$$\frac{\partial}{\partial t}h(u,t) + \sum_{i=1}^{n} f^i(t,u)\frac{\partial h(u,t)}{\partial u^i} + \sum_{i=1}^{n} \frac{\partial f^i(t,u)}{\partial u^i}h(u,t) = 0, \qquad (4.5.15)$$

in order to treat it as an extension of (4.5.9). We let $Z(u,t)$ be the solution of

$$Z_t(u,t) \;=\; \frac{\partial f^i(t,u)}{\partial u^i} Z(u,t) \qquad\qquad (4.5.16)$$

$$Z(u,0) \;=\; 1, \qquad\qquad (4.5.17)$$

and easily verify that the solution of (4.5.15), that is, of (4.5.7), then is determined by

$$h(Y(t,u),t)Z(u,t) = h(u,0) \text{ for all } t \in \mathbb{R}, u \in \mathbb{R}^n, \qquad (4.5.18)$$

that is, by an extension of (4.5.13). In order to get the C^1 regularity of h, we now need to assume that f is of class C^2, because (4.5.15) contains first derivatives of f. Assuming for simplicity that $h(u,t)$ has bounded support, we see from (4.5.7) directly that $\int_{\mathbb{R}^n} h(y,t)dy$ does not depend on t. Thus, the normalization (4.5.5) is preserved when we assume (4.5.6).
Transport equations in mathematical biology are systematically treated in [99].

We now also want to introduce diffusion effects into the dynamical law for the density function. We would like to discuss two approaches, a modern and a classical one. The reader can then decide for herself which one she prefers.

We first present the modern treatment which is based on the concept of white noise. Thus, we assume that the evolution equation (4.5.1) is subjected to white noise (we recall here the discussion of (4.2.53) above) of strength σ. Formally, one writes

$$u_t(t) = f(t, u) + \eta. \tag{4.5.19}$$

In order to understand what this means, we first put $f = 0$ and consider

$$u_t(t) = \eta. \tag{4.5.20}$$

This means that u randomly fluctuates in the sense that

$$u(t) = \int_0^t dw(\tau) + u(0) \tag{4.5.21}$$

where $w(t)$ is Brownian motion and the integral is a so-called Itô integral. Instead of explaining, however, what that means (see e.g. [69]), we rather state the corresponding equation for the probability density

$$\frac{\partial}{\partial t} h(u, t) = \frac{\sigma^2}{2} \Delta h(u, t) \tag{4.5.22}$$

where the Laplacian Δ operates on the u-variables, that is, $\Delta h(u, t) = \sum_{i=1}^{n} \frac{\partial^2}{(\partial u^i)^2} h(u, t)$. So, in contrast to the reaction-diffusion paradigm, here, the state variable is not diffusing in physical space, that is, for a variable $x \in \mathbb{R}^d$, but rather the state value is randomly fluctuating, leading to a diffusion for its density. Of course, (4.5.22) is the Fokker-Planck equation already studied above as a continuum limit of Brownian motion.

We can then combine the dynamical system (4.5.1) leading to the continuity equation (4.5.7) and the Fokker-Planck equation (4.5.22), to arrive at

$$\frac{\partial}{\partial t} h(u, t) = \frac{\sigma^2}{2} \Delta h(u, t) - \frac{\partial}{\partial u} (f(t, u) h(u, t)). \tag{4.5.23}$$

This is, of course, the same as (4.2.53) (except that there, we had only considered noise of strength 1), and we recall the discussion there. Again, a solution with Neumann boundary conditions satisfies (4.5.5) if it does so for $t = 0$.

The classical approach is based on the concept of a flux vector $J = (J^1, \ldots, J^n)$. The flux satisfies the conservation of mass law

$$\frac{\partial h}{\partial t} + \operatorname{div} J = 0 \tag{4.5.24}$$

where the divergence is defined as

$$\operatorname{div} J(u, t) := \sum_i \frac{\partial J^i}{\partial u^i}. \tag{4.5.25}$$

In general, the flux incorporates two contributions, a force term and a diffusion term. The force term is simply $f^i(t, u) h(u, t)$, as in (4.5.7). In other words, in the absence of diffusion, (4.5.24) reduces to (4.5.7). For the diffusion, on the other hand, we have Fick's law. This law says a diffusing substance moves from regions of higher to regions of lower concentration, at a rate proportional to the concentration gradient. That means, in the absence of an external force

$$J = -d\nabla h,$$
(4.5.26)

for some diffusion coefficient d, with the gradient

$$\nabla h(u, t) := (\frac{\partial h}{\partial u^1}, \ldots, \frac{\partial h}{\partial u^n}).$$
(4.5.27)

Since the operators involved satisfy

$$\mathrm{div}\nabla h = \Delta h,$$
(4.5.28)

as is readily verified, we obtain for the diffusion dynamics

$$\frac{\partial}{\partial t}h(u, t) = d\Delta h(u, t).$$
(4.5.29)

When we combine the effects of an external force and an intrinsic diffusion, the conservation of mass law (4.5.24) leads to the equation

$$\frac{\partial}{\partial t}h(u, t) = d\Delta h(u, t) - \sum_i \frac{\partial}{\partial u^i}(f^i(t, u)h(u, t)).$$
(4.5.30)

For $d = \frac{\sigma^2}{2}$, this is the same as (4.5.23).

The class of equations of the form (4.5.30) now can be used to model a variety of biological phenomena. An example that has also received much mathematical treatment is chemotaxis. This means the ability of organisms to react to concentration gradients of chemical substances by directed movement. These organisms can smell the molecules of particular substances, created by other members of their species and called pheromones in this context, and move up or down their concentration gradients. That is, these chemical substances can be attractive, like sex pheromones, or repulsive, like those used for the demarcation of territories. Here, we are not concerned with the molecular details of the reception mechanisms for such molecules, but rather with their effect on population densities.

We present here the Keller-Segel model for the chemotaxis of the slime mold *Dictyostelium discoideum*. These are colonies of single cell amoebae which produce the chemical substance cyclic-AMP that has an attractive effect on other amoebae in the colony. Thus, they move in the direction of the concentration gradient of this cyclic-AMP. The mathematical model then has to model the interplay between the population density h of the amoebae and the concentration c of cyclic-AMP as a function of space x and time t. A key quantity is the flux J of h. According to our preceding discussion, it contains the diffusive part of (4.5.26)

$$J_{\text{diffusion}} = -d\nabla h \tag{4.5.31}$$

and a chemotactic component

$$J_{\text{chemotaxis}} = h\chi(c)\nabla c, \tag{4.5.32}$$

for some function $\chi(c)$. Often, it is assumed that $\chi(c)$ is simply a constant χ_0. Whereas diffusion causes the minus sign in (4.5.31), the attractive nature of the pheromone leads to the positive sign in (4.5.32). With $J = J_{\text{diffusion}} + J_{\text{chemotaxis}}$, the conservation law (4.5.24) yields

$$\frac{\partial h(x,t)}{\partial t} = d\,\Delta h(x,t) - \text{div}(h(x,t)\chi(c(x,t))\nabla c(x,t)). \tag{4.5.33}$$

This now needs to be supplemented by an equation for the chemical concentration c. c is diffusing and decaying, but also produced by the amoebae cells, and therefore, there should be a reaction term proportional to the cell density h. Thus, the model postulates the equation

$$\frac{\partial c(x,t)}{\partial t} = \delta\Delta c(x,t) - \beta c(x,t) + \alpha h(x,t), \tag{4.5.34}$$

with positive constants δ, β, α. Since the chemical is diffusing faster than the cells, we should have $\delta > d$.

This model was introduced in [78, 79]. For extensive discussions and generalizations, see [92, 113]. For the more abstract PDE aspects, we refer to [99] and the references therein.

Exercises for This Chapter

1. Work out the details of the proof of the strong maximum principle for the Laplace operator on a graph.
2. Discuss the eigenvalue problem for the Laplace operator with periodic boundary conditions. **Hint:** Consider this problem as a version of Fourier analysis.
3. We now present some exercises on hyperbolic equations. The prototype is the wave equation. The wave equation in one space dimension is

$$u_{tt} - u_{xx} = 0 \quad \text{for } 0 < x < \pi, t > 0. \tag{4.5.35}$$

We impose initial data

$$u(x, 0) = \sum_{n=1}^{\infty} \alpha_n \sin nx, \quad u_t(x, 0) = \sum_{n=1}^{\infty} \beta_n \sin nx$$

and stipulate homogeneous boundary values

$$u(0, t) = u(\pi, t) = 0 \quad \text{for all } t > 0.$$

Compute the coefficients $\gamma_n(t)$ of the Fourier series representation

$$u(x, t) = \sum_{n=1}^{\infty} \gamma_n(t) \sin nx.$$

4. We next consider the first-order equation

$$v_t + c v_x = 0$$

for some function $v(x, t)$, $x, t \in \mathbb{R}$, where c is constant. Show that v is constant along any line

$$x - ct = \text{const} = \xi.$$

From this observation, conclude that the general solution for initial values $v(x, 0) = f(x)$ is

$$v(x, t) = f(\xi) = f(x - ct).$$

5. Observe that a solution of the wave equation (4.5.35) is given by

$$u(x, t) = \phi(x + t) + \psi(x - t) \tag{4.5.36}$$

for functions ϕ, ψ of class C^2. What is the connection of this observation with the preceding exercise?
6. We now turn to dynamical systems, that is, ODEs and their generalizations. Let $f : \mathbb{R} \to \mathbb{R}$ be Lipschitz, and consider the ODE

$$\dot{x}(t) = f(x(t)) \text{ for } t \geq 0, \text{ with } x(0) = x_0. \tag{4.5.37}$$

Show that if there are values $m < M$ with $f(m) = f(M) = 0$, $f(y) > 0$ for $m < y < M$, and if $m < x_0 < M$, then the solution of (4.5.37) exists for all $t > 0$, and

$$x_0 < x(t) < M \text{ for all } t > 0. \tag{4.5.38}$$

Show that this need no longer be true for delay-differential equations of the form

$$\dot{x}(t) = f(x(t - t_0)) \text{ or } \dot{x}(t) = f(x(t), x(t - t_0)) \text{ for } t \geq 0, \qquad (4.5.39)$$
$$\text{with } x(\tau) = x_\tau \text{ for } -t_0 \leq \tau \leq 0.$$

Hint: Look at the linear equation

$$\dot{x}(t) = -x(t - t_0) \qquad (4.5.40)$$

and observe that this equation admits periodic solutions.
Also, consider the delay version of the logistic equation (4.3.7),

$$\dot{x}(t) = x(t)(1 - x(t - t_0)), \text{ with } x(\tau) < 1 \text{ for } -t_0 \leq \tau < 0, x(0) = 1. \qquad (4.5.41)$$

Observe that in this case $\dot{x}(0) > 0$, and therefore $x(t) > 1$ for $0 < t < t_1$ for some $t_1 > 0$. When we take the maximal such t_1, then $x(t_1) = 1$ and $\dot{x}(t_1) < 0$ Conclude that then $x(t) < 1$ for $t_1 < t < t_2$. Again, take the maximal such t_2 and investigate what happens for $t > t_2$. Conclude that solutions of (4.5.41) may exhibit oscillatory behavior. We point out that in general, no analytical solutions for (4.5.41) are available, and this equation needs to be solved numerically. Try to do that, in order to get some understanding of the kind of behavior of delay-differential equations are capable of. For the theory of delay-differential equations, see [3], and for many biological applications, look into [92].

7. We now turn to problems with a some biological story behind them. In this exercise, we consider the Lotka-Volterra type model (4.3.60) with several species. In order to discuss a specific mathematical aspect, however, we assume that we have k predator and k prey species, that is, there numbers are the same (formally, of course, this can be achieved, by simply dividing up some species into different classes, in order to inflate species numbers). The prey species numbers are given by $y^1, \ldots, y^k$, and those of the predator species by $z^{k+1}, \ldots, z^{2k}$. The dynamics are

$$\dot{y}^j = y^j\left(a_j - \sum_{\ell=k+1, \ldots, 2k} b_{j\ell} z^\ell\right) \quad \text{for } j = 1, \ldots, k$$
$$\dot{z}^\ell = z^\ell\left(a_\ell - \sum_{j=1, \ldots, k} b_{j\ell} y^j\right) \qquad \text{for } \ell = k + 1, \ldots, 2k,$$

where the coefficients satisfy

$$a_j > 0 \quad \text{for } j = 1, \ldots, k, \qquad\qquad a_\ell < 0 \text{ for } \ell = k + 1, \ldots, 2k,$$
$$b_{j\ell} < 0, \quad b_{\ell j} > 0 \quad \text{for } j = 1, \ldots, k, \ \ell = k + 1, \ldots 2k.$$

Find a nontrivial fixed point. As in the two species model, this fixed point corresponds to a steady state where the population sizes all stay constant. Linearize the dynamics about that fixed point. Observe that the matrix A appearing in that linearization has trace $A = 0$. Conclude that therefore its eigenvalues $\lambda_1, \ldots, \lambda_{2k}$ satisfy $\sum_{i=1}^{2k} \lambda_i = 0$. Since A is real, the eigenvalues are either real or occur in complex conjugate pairs. Conclude that either all of them have vanishing real part, in which case the steady state is neutrally stable as in the two-species case. Or at least one eigenvalue has to have a positive real part, in which case the steady state is unstable. Thus, any small perturbation of the coefficients of the system that makes some of the real parts of the eigenvalues nonzero destabilizes the steady state. You may want to conclude that the model is too simple to realistically capture the population dynamics of several interacting species.

8. We next look at a Lotka-Volterra model in the case of symbiosis,

$$
\begin{aligned}
\dot{x}^1 &= x^1(a_1 + b_{12}x^2) \\
\dot{x}^2 &= x^2(a_2 + b_{21}x^1),
\end{aligned}
\tag{4.5.42}
$$

with $a_1 > 0$, $b_{12} > 0$, $b_{21} > 0$, while a_2 is arbitrary. First observe that in this model, the population sizes will go to ∞ if they start with positive values. Therefore, we modify the model by growth restricting terms, as in (4.3.62). After some coefficient normalization, the model then becomes

$$
\begin{aligned}
\dot{x}^1 &= x^1(1 - x^1 + b_{12}x^2) \\
\dot{x}^2 &= ax^2(1 - x^2 + b_{21}x^1),
\end{aligned}
\tag{4.5.43}
$$

with $a, b_{12}, b_{21} > 0$. The steady states of this system are

$$
(0,0), \quad (1,0), \quad (0,1), \quad \left(\frac{1+b_{12}}{1-b_{12}b_{21}}, \frac{1+b_{21}}{1-b_{12}b_{21}}\right).
\tag{4.5.44}
$$

Show that the first three are unstable, while the last one is only positive if $1 - b_{12}b_{21} > 0$. Show that in this case, that equilibrium is stable. Conclude that when $1 - b_{12}b_{21} < 0$ populations with positive initial sizes will tend to ∞ whereas when $1 - b_{12}b_{21} > 0$, the population sizes will converge to the last fixed point in (4.5.44).

9. We conclude with some exercises on the Turing mechanism. First, determine the Turing spaces for the Gierer-Meinhardt and Thomas systems.

10. Carry out the analysis of the Turing mechanism for periodic boundary conditions.

11. This exercise discusses the Brusselator model, introduced by Lefever, Nicolis, and Prigogine (in Brussels) as a toy model for pattern formation in reaction-diffusion systems. The model is based on the system of chemical reactions

$$
\begin{aligned}
A &\to X; \quad B + X \to Y \\
2X + Y &\to 3X; \quad X \to E.
\end{aligned}
\tag{4.5.45}
$$

Note the similarity with the Schnakenberg reactions (4.4.85). The concentrations of A, B, and E are kept constant and therefore can be used as control parameters. After rescaling, the concentrations of X and Y are expressed by variables u, v satisfying the system

$$u_t \;=\; \Delta u + a - (b+1)u + u^2 v$$
$$v_t \;=\; d\Delta v + bu - u^2 v. \tag{4.5.46}$$

Here, a and b are constants (corresponding to the concentrations of A and B), and we shall use b as a bifurcation parameter. As in Sect. 4.4.4, we assume that v diffuses faster than u, that is, $d > 1$. Observe that this system admits the steady state

$$u_0 = a, \quad v_0 = \frac{b}{a}. \tag{4.5.47}$$

First, carry out a linear stability in the absence of diffusion (i.e., put $\Delta u = \Delta v = 0$ for the moment), by linearizing (4.5.46) at the steady state (4.5.47). Observe that one of the eigenvalues of the linearization matrix changes its real part from negative to positive values when $b = b_{Hopf} = 1 + a^2$. Conclude that at this parameter value, a Hopf bifurcation takes place, that is, a transition from a uniform steady state to limit cycle oscillations.

Next, expand u and v in terms of eigenfunctions u_k as in (4.4.96) and look for solutions of the form (4.4.97) and conclude that there is a Turing instability at

$$b = b_{Turing} = \left(1 + \frac{a}{\sqrt{d}}\right)^2, \tag{4.5.48}$$

with frequency

$$\lambda_{Turing} = \sqrt{\frac{a}{\sqrt{d}}}. \tag{4.5.49}$$

(In contrast to the situation analyzed in Sect. 4.4.4, here the critical value depends only on the kinetic parameters, but not on the underlying geometry.)

The constant solution u_0, v_0 that bifurcates into a limit cycle at b_{Hopf} then corresponds to the frequency 0. For which ranges of $\frac{d_u}{d_v}$ does the Hopf bifurcation occur before the Turing instability, i.e., when is $b_{Hopf} < b_{Turing}$, and when is it the other way around? Note that the Turing instability is only of interest when it occurs before the Hopf bifurcation as after the latter, the equilibrium (4.5.47) is already unstable.

References for this exercise are [120], Sect. 14.3, and [98].

Chapter 5
Optimization

Abstract
Questions:

- How do we best distribute our finite resources among different tasks?
- Isn't sexual reproduction wasteful? Why are there males?
- What are the criteria for the optimality of a functional relationship?

Biological evolution is about getting an advantage by becoming better than others. That is, optimization problems arise and should be solved. A key problem is the best allocation of finite resources to different tasks. We develop this in a simple setting. As an application, we can explain why sexual reproduction is prevalent inspite of its apparent shortcomings. We also introduce the calculus of variations as the mathematical theory for the optimization of functional relationships.

5.1 Optimization of Resource Allocation

The basic fitness function in evolutionary biology is reproductive success, that is, the number of viable offspring, possibly counted over several generations. An individual or lineage that produces more viable offspring than others is genetically better represented in the next generation. For a deeper discussion of this issue, which turns out to be not as simple as it may appear, we refer to [66]. Here, we simply analyze the optimization of reproductive strategies in an abstract setting. We try to identify the best possible strategies, since those provide the limit for improvements. The problems involved are often allocation problems. Organisms often have to decide about the allocation of some limited resource among several potential uses. Evolutionary competition leads to optimization pressures for allocation strategies. This can be complicated by the fact that the rewards for each allocation decision may depend on the allocation strategies of other individuals. Thus although evolutionary optimization typically occurs at the level of the individual, the collective effect of individual optimization changes rewards. This then leads to feedback effects between

J. Jost, *Mathematical Methods in Biology and Neurobiology*, Universitext
DOI: 10.1007/978-1-4471-6353-4_5, © Springer-Verlag London 2014

individual optimization and population-level patterns. Therefore, in this section, we develop the mathematical framework both for individual allocation optimization and for the analysis of population feedback effects and the resulting population dynamics.

In this section, which is based on [75], we develop the mathematical framework for such allocation problems.

5.1.1 Cost and Reward

We consider a state space that is defined by certain degrees of freedom, represented by independent variables $x_1, x_2, \ldots, x_n$. x_i quantifies the investment allocated to option i. The vector $x = (x_1, \ldots, x_n)$ thus represents a strategy, that is, a decision about the allocation of the available resources to the possible investment options. The state space then consists of all possible resource allocation strategies. The x_i will therefore satisfy some constraints like $x_i \geq 0$ for all i and some relations like $g_j(x_1, \ldots, x_n) = c_j$ for $j = 1, 2, \ldots$, for example $\sum_i x_i = \text{const.}$, that is, the total amount of available resources is fixed. Unless explicitly stated otherwise, these constraints and other functions occuring in the sequel will be assumed to be sufficiently differentiable. If those relations are independent they define some manifold M, and the independent variables then can be thought of as varying in M. More precisely, we shall assume that the independent variables are constrained to be nonnegative and to satisfy the *cost constraint*

$$C(x_1, \ldots, x_n) \leq c, \tag{5.1.1}$$

where $C(x)$ is the total cost of the investment strategy x, and c is the highest total cost an individual can spend.

The individuals are assumed to try to *maximize their reward or gain* (we shall use these words equivalently) under these constraints

$$R(x_1, \ldots, x_n) \to \max. \tag{5.1.2}$$

We shall also assume that both C and R are monotonically increasing functions of their arguments. This is, of course, biologically plausible. For reward optimization, the cost constraint thus becomes the equality

$$C(x_1, \ldots, x_n) = c. \tag{5.1.3}$$

A maximum that is achieved at an interior point, i.e. at a point where all the x_i are positive, has to satisfy the Lagrange multiplier rule

$$\frac{\partial R(x_1, \ldots, x_n)}{\partial x_i} + \lambda \frac{\partial C(x_1, \ldots, x_n)}{\partial x_i} = 0 \text{ for } i = 1, \ldots, n \tag{5.1.4}$$

for some real λ. In particular, this implies

$$\frac{\partial_i R(x_1, ..., x_n)}{\partial_i C(x_1, ..., x_n)} = \frac{\partial_j R(x_1, ..., x_n)}{\partial_j C(x_1, ..., x_n)} \quad \text{for all } i, j \tag{5.1.5}$$

(using the abbreviation $\partial_i = \frac{\partial}{\partial x_i}$) or, expressed verbally, that the ratio

$$\frac{\text{marginal gain}}{\text{marginal cost}} \tag{5.1.6}$$

is the same for all x_i at equilibrium. If the maximum is achieved at some boundary point where some of the x_i vanish, then (5.1.5) continues to hold for those indices i, j for which $x_i > 0, x_j > 0$ while for the indices α with $x_\alpha = 0$, we get

$$\frac{\partial_\alpha R(x_1, ..., x_n)}{\partial_\alpha C(x_1, ..., x_n)} \leq \frac{\partial_i R(x_1, ..., x_n)}{\partial_i C(x_1, ..., x_n)}. \tag{5.1.7}$$

Thus, the quotient of marginal gain and marginal cost in the constrained directions cannot exceed that in the unconstrained directions. This means that by increasing x_α above 0, one cannot gain more than one looses when the cost condition necessitates a corresponding reduction of some of the unconstrained variables x_i.

Such a condition on first derivatives is a necessary condition for a maximum; for a sufficient condition, we need a condition involving second derivatives, that is, some kind of relative convexity or concavity of the functions C and R at the considered point.

The optimum need not be unique, and in the sequel, we shall discuss some cases with two stable optima.

We also observe that cost and reward are dual to each other. Instead of maximizing the reward while keeping the costs constant, we could as well minimize the costs while keeping the reward fixed.

The preceding can, of course, also be explained in geometric terms. We consider the cost level $C(x) = c$ and look for points on this level set where it touches or meets the highest reward level set. This can happen at an interior point, i.e. where all the x_i are positive, or at a boundary point where some of them vanish. If that highest attainable reward level set meets the cost level at more than one point, the optimum is nonunique. The optimal reward level touches the cost level set at an interior point or meets it at a boundary point, but cannot intersect it, because otherwise in the vicinity of such an intersection, we would also find higher reward level sets that also intersect the given cost level. That would mean, however, that on the cost level a higher reward could be realized, in contrast to the assumption that we are at an optimum.

By a coordinate transformation we may, in fact, assume that the cost levels are flat (this simply means that all quantities are expressed in units of cost). In fact, we can measure all items in cost units normalized so that each unit of each of the independent variables $x_1, ..., x_n$ costs the same. This yields

$$x_1 + \ldots + x_n = C = \text{const} \tag{5.1.8}$$

We shall make this assumption for the rest of this section.

In the case of two independent variables, the cost levels are then straight lines (of slope -1 with the indicated normalization).

5.1.2 Reward Functions and Strategy Types

With the above normalization (5.1.8) that the cost levels are flat, we shall now discuss qualitative types of reward functions.

Definition 5.1.1. The items $i = 1, \ldots, n$ are called *substitutable* if the reward function is of the form

$$R(x_1, \ldots, x_n) = r_1 x_1 + \ldots + r_n x_n \tag{5.1.9}$$

with positive constant coefficients $r_1, \ldots, r_n$.

The items are called *complementary* if the reward function is of the form

$$R(x_1, \ldots, x_n) = \min(s_1 x_1, \ldots, s_n x_n), \tag{5.1.10}$$

again with positive coefficients s_i.

The items are called *mutually enhancing* or *mutually conflicting* if the reward function has convex or concave resp. level sets.

An example of a reward function with mutually enhancing items is

$$R(x_1, \ldots, x_n) = x_1^{\alpha_1} \ldots x_n^{\alpha_n} \tag{5.1.11}$$

with positive exponents α_i. A mutually conflicting example is

$$R(x_1, \ldots, x_n) = v_1 x_1^{\beta_1} + \ldots + v_n x_n^{\beta_n} \tag{5.1.12}$$

with positive coefficients v_i and exponents $\beta_i \leq 1$.

A biological example of mutually enhancing rewards is given by the allocation of life time to the growth period and the reproductive period. In annual insects that undergo metamorphosis as part of the life cycle, the larval stage is the growth period whereas in the adult stage, they concentrate on reproduction. The cost constraint here is the fixed life span of one season.

The qualitative features of the reward function will constrain the position of the optimum. If the items are mutually conflicting, i.e. if the reward levels are concave, then an optimum can only be achieved at a boundary point where all but one of the x_i vanish. Such an optimum need not be unique. Below, we shall discuss a scenario where population effects will make the optimum nonunique at the individual level. In that scenario, the items will be complementary at the population level, but mutually conflicting at the individual one. If the items are mutually enhancing, then an optimum can be achieved either at an interior or a boundary point. If the items are complementary, then the unique optimum is always achieved at that interior point where

$$s_1 x_1 = s_2 x_2 = \ldots = s_n x_n \qquad (5.1.13)$$

as otherwise the agent would waste resources without a reward increase. Below, we shall interpret (5.1.13) as a *balance equation*. Finally, if the items are substitutable and the largest one among the r_i in (5.1.9) is unique, then the optimum is realized at that boundary point on the cost level where only that product x_j is nonzero that comes with the largest coefficient r_j. If some of the largest coefficients r_j are equal, then the optimum is degenerate in the sense that the agent can shift resources between all the corresponding x_j while keeping cost and reward the same. Although this seems to be a very special situation, we shall encounter it below in our discussion of the optimization of sex ratios.

Of course, the above classification of reward functions is not complete, and, for example, many reward functions will have neither convex nor concave level sets, but rather level sets that change type, i.e., are convex in certain regions and concave in others. In fact, if we leave degenerate cases (e.g. complete substitutability) aside, only for such reward functions there may exist more than one interior optimum. It is also possible that an interior and a boundary optimum coexist.

Definition 5.1.2. A strategy is called a *generalist* strategy if all the x_i are positive. It is called a *specialist* strategy if all but one of the x_i vanish.

Below, we shall encounter reward functions that are complementary at the population level, but possibly substitutable or even mutually conflicting at the individual level.

5.1.3 Complementarity

What we want to develop and apply the mathematical framework for is the interaction between individual optimization and population effects. A typical situation is specialization in the situation of complementary productions. We shall therefore analyze this issue and obtain an abstract version of the Fisher equilibrium condition [45]. We shall apply this below to the issue of sexual reproduction which is a typical case of such complementarity. In that context, the units can either be gametes (of male or female type) or (male or female) offspring. When we then speak of a population, in the first case we shall mean a collection of gametes. The reward to be optimized then is the number of offspring or, equivalently, of successful matings in the case of gametes. In the case of unisexual offspring, the reward is the number of grand offspring (we shall see below why we should consider second instead of first generation offspring). The crucial feature is the complementarity, stemming from the fact that it always requires the fusion of two units of opposite types, called male and female for concreteness, to generate a reward unit. For instance, an individual can be female, that is, produce female gametes, but it then requires a male mating partner.

Therefore, we assume that a balance equation holds at the population level. If the population consists of n members, and if the reproductive success of each individual i through male/female units is $R_{m,i}$ and $R_{f,i}$, resp., we then have the balance condition

$$\sum_{i=1}^{n} R_{m,i} = \sum_{i=1}^{n} R_{f,i}. \tag{5.1.14}$$

If m_i and f_i denote the numbers of male and female units, resp., produced by individual i, and if its reward and cost are $R(m_i, f_i)$ and $C(m_i, f_i)$, resp., and if a subscript 1 or 2 denotes a partial derivative w.r.t. m_i or f_i, resp., then from the previous section, we obtain the equilibrium condition

$$\frac{\partial_1 R(m_i, f_i)}{\partial_1 C(m_i, f_i)} = \frac{\partial_2 R(m_i, f_i)}{\partial_2 C(m_i, f_i)}, \tag{5.1.15}$$

at least if that equilibrium is attained at an interior point of the positive quadrant, i.e. at a point where m_i and f_i are both positive.

Let us understand the meaning of such an equilibrium in our biological example of sexual reproduction. In the gamete allocation case, an interior equilibrium point corresponds to hermaphroditic behavior. This means that the corresponding individual is bisexual or monoecious, that is, produces both male and female gametes. The opposite strategy where only gametes of a single type are produced is called dioecious. This would correspond to a boundary optimum. Because of the complementarity, we then need two boundary optima, one representing a pure male and the other a pure female strategy. Population affects will then keep these two optima in balance, as will be discussed below. Even without requiring complementarity, we already obtain the Eqs. (5.1.5, 5.1.6). Let us briefly the example of linear cost and reward structures. The reward of agent i is then

$$R(i) = m_i r_m(m_i, f_i) + f_i r_f(m_i, f_i), \tag{5.1.16}$$

and its cost is

$$C(i) = c_m(m_i, f_i)m_i + c_f(m_i, f_i)f_i. \tag{5.1.17}$$

The equilibrium condition then becomes

$$\frac{r_m + m_i \partial_1 r_m + f_i \partial_1 r_f}{c_m + m_i \partial_1 c_m + f_i \partial_1 c_f} = \frac{r_f + m_i \partial_2 r_m + f_i \partial_2 r_f}{c_f + m_i \partial_2 c_m + f_i \partial_2 c_f}, \tag{5.1.18}$$

If the coefficients c_m, c_f, r_m, r_f are all constant, i.e independent of the numbers of units produced, (5.1.18) reduces to

$$\frac{r_m}{c_m} = \frac{r_f}{c_f} \tag{5.1.19}$$

i.e. that the cost of a *successful* unit is the same for either type. This equilibrium is stable; if for example $\frac{c_m}{r_m}$ were smaller than $\frac{c_f}{r_f}$ then i could increase $R(i)$ by increasing m_i (and correspondingly decreasing f_i in order to satisfy the energy constraint), and vice versa.

In order to analyze the interplay between the individual and the population levels, we now invoke the balance Eq. (5.1.14). The balance equation says

$$r_m(m_p, f_p)m_p = r_f(m_p, f_p)f_p. \tag{5.1.20}$$

It is important to note that the subscript p now refers to the numbers produced by the population as a whole. Likewise, r_m, r_f now denote the average success rates at the population level.

We insert this relation into the equilibrium condition (5.1.19) (assuming constant coefficients) and obtain

$$c_m m_p = c_f f_p. \tag{5.1.21}$$

Thus, from the balance equation we conclude that the investments into male and into female reproduction have to be or turn out to be equal at the population level. Thus, the complementarity at the population level lets individual optimization lead to such a global equilibrium. And this equilibrium is stable; if, for instance, $c_m m_p$ is smaller than $c_f f_p$, then it becomes advantageous for an individual i to increase m_i at the expense of f_i.

5.1.4 Dynamical Interaction Between Individual Strategies and Population Effects

We want to understand how the composition of the population can determine the reward structure for its individual members. Individual agents can vary their own behavior, thereby increasing their own award, but without causing a discernible effect at the population level. If, however, the population is homogeneous, then all agents share the same reward structure, and so, when the optimum is unique, then all agents will attempt to modify their strategies towards that individual optimum. When that optimum, however, sensitively depends on the collective behavior of the population, then the systematic change of all individual strategies will change this global optimum in turn. Thus, the collective action will affect each individual's situation, and, in particular, the parameters on which the optimization is based may not remain constant. In many cases, there is a negative feedback between the individual and the population level reward structure, for instance if the reward structure at the population level is complementary. In order to understand this better, we now discuss a model where the individual and the population dynamics occur on the same time scale. (In other situations, the population effects might show themselves more slowly, as in many ecological scenarios.)

We consider a population of N individuals. The population level quantities are the sums of the individual terms, i.e.,

$$m_p = \sum_{j=1}^{N} m_j, \ \ f_p = \sum_{j=1}^{N} f_j.$$

We assume that the population is homogeneous so that every agent i has the same contribution, i.e.,

$$m_i = \frac{1}{N} \sum m_j, \ \ f_i = \frac{1}{N} \sum f_j. \tag{5.1.22}$$

We now consider the dynamical setting where the agents find themselves in a non-optimal situation and therefore perform a gradient ascent for the function $R(i)$ under their cost constraint. Let t be time, and denote a derivative w.r.t. t by a dot. We also use the superscript T for the projection of the gradient onto the level set of the cost function. When the reward coefficients r_m, r_f and the cost coefficients c_m, c_f are constant, the projected gradient of R is

$$\partial_1^T R = r_m - \frac{r_m c_m + r_f c_f}{c_m^2 + c_f^2} c_m, \ \ \partial_2^T R = r_f - \frac{r_m c_m + r_f c_f}{c_m^2 + c_f^2} c_f. \tag{5.1.23}$$

The cost constrained optimization then leads to the dynamical system

$$\dot{m}_i = \partial_1^T R(i), \ \ \dot{f}_i = \partial_2^T R(i) \tag{5.1.24}$$

The reward of i depends both on the individual strategy and on the population level quantities,

$$R(i) = R(m_p, f_p, m_i, f_i). \tag{5.1.25}$$

With this, the dynamics of (5.1.24) leads to

$$\dot{R}(i) = \frac{\partial R}{\partial m_p} N \dot{m}_i + \frac{\partial R}{\partial f_p} N \dot{f}_i + \partial_1 R \, \dot{m}_i + \partial_2 R \, \dot{f}_i$$

$$= (N \frac{\partial R}{\partial m_p} + \partial_1 R) \partial_1^T R + (N \frac{\partial R}{\partial f_p} + \partial_2 R) \partial_2^T R. \tag{5.1.26}$$

Typically, the reward of individual i will be positively correlated with her own production, but inversely related to the competing production from the rest of the population, and again positively correlated with the complementary production of the rest of the population. Let us consider the following example

$$r_m = \min(\frac{f_p}{m_p}, 1), \quad r_f = \min(\frac{m_p}{f_p}, 1)$$

and

$$R(i) = r_m m_i + r_f f_i, \tag{5.1.27}$$

Moreover, let us assume $m_p \leq f_p$. In this case, r_f will remain constant. The dynamics of $R(i)$ then is

$$\dot{R}(i) = \dot{r}_m m_i + r_m \dot{m}_i + r_f \dot{f}_i \quad \text{since } r_f \text{ is constant}$$

$$= (\frac{N}{\sum f_j} \partial_2^T R - \frac{N}{\sum m_j} \partial_1^T R) r_m m_i + r_m \partial_1^T R + \partial_2^T R$$

since all N agents follow the same dynamics (5.1.24)

$$= 2\partial_2^T R \quad \text{since } N m_i = \sum m_j \text{ and } r_m = \frac{\sum f_j}{\sum m_j}. \tag{5.1.28}$$

When $\partial_2^T R \leq 0$, all agents will therefore increase their female production, leading to a gain of $\partial_2^T R$ for each of them, while the individual loss from the decrease of the male production is more than offset by the gain resulting from the population effect that all other agents also increase their female production, thereby increasing everybody's reward from male production. If, however, $\partial_2^T R < 0$, for example because c_f is large, then all agents will decrease their female production, and for each agent, the resulting increase in male production by the rest of the population also makes its male gametes less successful so that a loss for everybody results.
In any case, the dynamics will go on until

$$\frac{r_m}{c_m} = \frac{r_f}{c_f} = \frac{r_m c_m + r_f c_f}{c_m^2 + c_f^2}. \tag{5.1.29}$$

This is again the Fisher equilibrium condition that we have seen earlier, see e.g. (5.1.19).

5.1.5 Generalizations

In the preceding, we have investigated situations with only two independent variables, the male and female units m_i, f_i. We now generalize this by enlarging the strategy space of the individuals, or in biological terminology, by including other strategic choices that may enhance the reproductive success, like mating effort or parental care. We shall assume that such factors can be varied independently of the investments into male or female units. We can then discuss the possible effects of a trade-off between the efforts for the production of gametes and for achieving matings or for investments into parental care.

So, for a formal analysis we assume that in addition to m_i and f_i, we have another factor y_i that enters into the cost function

$$C(m_i, f_i, y_i) \tag{5.1.30}$$

and the reward function

$$R(m_i, f_i, y_i). \tag{5.1.31}$$

∂_3 will denote a derivative w.r.t. y. At an interior equilibrium under the cost constraint

$$C(m_i, f_i, y_i) = c_i, \tag{5.1.32}$$

we then have

$$\frac{\partial_1 R(m_i, f_i, y_i)}{\partial_1 C(m_i, f_i, y_i)} = \frac{\partial_2 R(m_i, f_i, y_i)}{\partial_2 C(m_i, f_i, y_i)} = \frac{\partial_3 R(m_i, f_i, y_i)}{\partial_3 C(m_i, f_i, y_i)}. \tag{5.1.33}$$

The Eqs. (5.1.32, 5.1.33) can be solved w.r.t. the variables m_i, f_i, y_i when the corresponding functional determinant does not vanish.

As an example, let us assume a linear cost structure with constant coefficients

$$C(i) = C(m_i, f_i, y_i) = c_m m_i + c_f f_i + c_y y_i \tag{5.1.34}$$

and a reward function that is linear in m_i, f_i

$$R(i) = m_i r_m(y_i) + f_i r_f(y_i). \tag{5.1.35}$$

We then obtain

$$c_m m_i = c_y \frac{r_m}{\partial_3 r_m} \tag{5.1.36}$$

and if we further assume that r_m is proportional to y_i, we get the optimality condition

$$c_m m_i = c_y y_i. \tag{5.1.37}$$

The individual i thus invests the same amount into male units as into the new factor. If r_f is likewise proportional to y_i, these quantities then in turn would also have to equal $c_f f_i$. Thus, we obtain (5.1.21) without having to appeal to the balance equation at the population level.

If r_m is more sensitive to y_i than r_f, then m_i will be more sensitive to a change in the cost c_y, and thus, an increase in c_y will decrease the ratio $\frac{m_i}{f_i}$.

Another biologically interesting example is that y_i is complementary to m_i and f_i already at the individual level. This means that for successful reproduction, any investment into male or female units has to be matched by some investment into the new factor. This leads to a reward function of the form

$$R(i) = \min(r_m y_i, m_i) + \min(r_f y_i, f_i). \tag{5.1.38}$$

For example, the coefficients could be

$$r_m = \rho \min(\frac{f_p}{m_p}, 1), \quad r_f = \rho \min(\frac{m_p}{f_p}, 1) \tag{5.1.39}$$

for some constant ρ, i then reaches its optimum if under the cost constraint (5.1.32)

$$r_m y_i = m_i \geq f_i \text{ or } r_f y_i = f_i \geq m_i \tag{5.1.40}$$

as otherwise it would waste some of its units produced. When, for instance, the first alternative is realized, then i's reward is

$$R(i) = r_m y_i + f_i \tag{5.1.41}$$

with the cost structure

$$c_f f_i + (r_m c_m + c_y) y_i = c_i. \tag{5.1.42}$$

Our above equilibrium analysis (5.1.19), with y_i in place of m_i, yields the equilibrium condition

$$\frac{r_m}{r_m c_m + c_y} = \frac{1}{c_f}. \tag{5.1.43}$$

Since we assume that all coefficients are constant and fixed, this condition may or may not hold at the given levels of m_p, f_p. When the left hand side of (5.1.43) is larger than the right hand side, the agent i will shift effort from the production of f_i to the production of y_i. If we have a homogeneous population of hermaphrodites, all with the same production of m_i, f_i, y_i, then all agents will move in the same direction, thereby decreasing r_m until equality holds in (5.1.43). This occurs for example if c_m is very small compared to the other cost coefficients, i.e. if the production of male units is very cheap, and if c_y in turn is considerably smaller than c_f. Then one might simply merge the costs for male units and for the new factor and redefine the male costs as those for the production of male units and for y together. This changes, if, while c_m still is very small, c_y now is much larger than c_f. Then the right hand side of (5.1.43) is the larger one, and the agents will shift their efforts to the production of f_i, until

$$r_f y_i = f_i. \qquad (5.1.44)$$

A further increase of f_i then will no longer lead to any further gain. Thus, high costs for an additional factor entering into the reward function complementary to the others produce stable hermaphroditic behavior. The reason is that any investment into the new expensive factor will benefit both types of units, and so increasing the production of both may also increase the benefits from any investment into that factor. The biological effect could be a more intense competition among males than among females for matings.

The case just treated may describe a situation where the factor y stands for mating encounters. Thus, mating costs that are higher than the production costs for gametes may support hermaphroditism. A prominent biological are plants that depend on insect pollination. In contrast, if the production of female gametes is most expensive, the mating costs may be carried by the males in a dioecious, i.e., bisexual population, because additional matings are more beneficial for males than for females.

For the biological example of parental investment, we return to the example where the reward coefficients are proportional to the quantity of the new factor. As the female reward might be more sensitive to parental care, and so, one might expect the females to invest more into that factor than the males. This, in fact, is the case in many biological examples, but there also exist cases of higher male parental investment.

When there are more factors, then complementarity might also develop between two of them, like e.g. between mating efforts and parental care. This produces a third level of complementarity, concerning individual behavior. The Fisher equilibrium again applies to derive an equilibrium condition for investments at the behavioral level. A specialized strategy could, for instance, consist in the males investing exclusively into mating efforts and the females only into parental care. Coexistence of pure and mixed strategies is also possible in principle and realized in many mammalian populations.

The alert reader should be able to find many more examples where such an equilibrium can be applied.

5.1.6 Why do We Have Sex?

The preceding can easily and naturally explain sexual reproduction. In fact, it can explain it at two different levels, as our considerations can be applied to both individuals and gametes. First of all, we can see Fisher's answer [45] to the question why the sex rations in natural populations are typically balanced, that is, there are about equal numbers of males and females. One may wonder why this is so when the principle of evolution is to simply produce as much viable offspring as possible. For that, it would not matter whether that offspring is female or male. However, when we consider the number of grand offspring, that is, second instead of first generation offspring, as the reward criterion, then we can see the effect of the above complementarity at the population level. If for instance, the population is predominantly female, then a male has a much higher expected number of offspring than a female, and conversely, simply because in a sexually reproducing population every individual needs to have both a mother and a father. Thus, in a predominantly female population, it would be better to produce male offspring. This shows that a balanced sex ratio constitutes a stable equilibrium.

In fact, this reasoning is somewhat of a simplification of our formal analysis. That analysis tells us more precisely that the effort in producing offspring of the two sexes should be balanced in terms of cost units, that is, in terms of energy spent. For instance, let us assume that, for instance in some insect species, five times more energy is required to produce a female than a male. Then the above analysis yields an equilibrium sex ratio of five males for one female. In most vertebrate species, however, there are no such drastic energy differences, and so, their sex ratios are about equal. In particular, we here see the difference between optimization at the individual and at the population level. At the population level, it would be best to have as many females as possible with a number of males just sufficient to fertilize all the females. That would yield the highest growth rate for the population because males can potentially sire much more offspring than females. However, in such a situation where the male reproductive chances of males would be so much higher than those of females, for every individual couple the best strategy would be to produce only male offspring. This would continue to hold until the sex ratio is balanced. Thus, individual optimization is not optimal for the population as a whole.

Our analysis can also explain sexual reproduction at a more fundamental level, that is, answer the question why there is sexual reproduction at all, instead of simple reproduction by cell division as in bacteria or by simple cloning. There exist many proposals in the biological literature for explaining the origin and persistence of sexual recombination. Surveys and critical discussions can be found in [85, 46]. Some theories are based on the advantage of sexual recombination at the population level, like the fixation of rare beneficial mutations, or conversely, the elimination of deleterious mutations. Other theories are built upon the advantages of genetic diversity in heterogeneous habitats or under pressure from parasites. Sometimes, sexual reproduction is considered as a puzzle because a population consisting of individuals that can reproduce without mating partners by self-fertilization could produce twice as much offspring as a population where only the females can lay eggs or bear babies. In the biological discussion, however, there often is some confusion about the issues involved. In fact, to understand the problem of sexual reproduction, we should clearly separate two issues. One is genetic recombination, and the other is sexual specialiazation.

So, let us start with genetic recombination, or more basically, with biological variation. We shall argue on the basis of a somewhat coarse summary of the theory of evolution. Evolution is concerned with finding ever better adapted organisms as the fitter outcompete the less fit. This is a search, however, where the answer is not known beforehand. This means that new possibilities need to be tried out so that by chance a better solution can be stumbled upon. The easiest possibility whereby such a variation can be produced is through mutations. This simply means that the offspring is not genetically identical to its parent, but carries some genetic changes. The problem, however, is that most mutations are disadvantageous, deleterious or even lethal for its carrier, and so, generating too many mutations is in general not a successful evolutionary strategy. After all, an organism that has survived to maturity in general already is well adapted and difficult to improve upon through such variations. Thus, biologically mutation rates are usually very small (with exceptions that need not concern us at this point). There is a solution to the dilemma that on one hand, variations are needed in order to explore new possibilities and that on the other hand random mutations have a high probability of carrying negative in place of positive effects. This solution is recombination. Recombination with partners that have demonstrated their abilities by survival gives access to variations of offspring that have already been tested. The natural candidates for such successful mutations can be found in other individuals in the same ecological niche. Through recombination, an individual can incorporate genetic material from another individual in its offspring in place of completely random mutations. As this is a symmetric situation between two individuals, one solution is that two individuals exchange half of their genomes. Thus, each individual then is only genetically represented in its offspring with a contribution of one half, but this is compensated by the fact that she is also represented to the same extent in the offspring of the partner. This can be a completely symmetric situation, where neither parent invests more than the other into the offspring. So far, this is neutral at the population level, in the sense that the average number of offspring per individual is the same as before. In particular, the reduction of the offspring number by one half has not yet occurred. So, why does it occur, that is, why is there an equilibrium where only one half of the population, the females, produces offspring, whereas the other half, the males, seem to simply waste their energy in seeking matings?

The answer is that, in this scenario, optimization through specialization can take place. This is because the success of reproduction now depends on two different factors that may be impossible to maximize simultaneously. On one hand, a gamete should carry plentiful nutrients to nourish the developing offspring, and on the other hand, high mobility of gametes is needed so that they can find and fertilize each other. This creates the opportunity for optimization of gametes through differgent specializations. The result of such spezialiations is anisogamy, that is, the difference between sperm and egg. So far, this is a specialization of gamete types, but not necessarily of individuals. Individuals could still be hermaphroditic, that is, carry both types of gametes. Again, in this case, the number of offspring would not be reduced compared with asexual reproduction, as each individual produces offspring from its eggs. The situation is still symmetric between the mating partners. When, however, two different types of gametes exist, with different factors increasing the success of each, this can in turn induce another round of optimization through specialization, this time at the level of the organism. Instead of being hermaphrodites, that is, generalists that pursue reproduction through both eggs and sperm, they may rather specialize in reproduction exclusively through sperm or through eggs, and become males and females. In principle, in a population, both generalists and specialized genders could coexist. We have developed a variational scheme to understand how individual optimization can lead to the emergence of two different mating types, female and male, via specialization. This specialization then is stabilized by the dynamics at the population level. Now we have a twofold cost of sexual reproduction because if the females developed a method of self-fertilization, the males would become superfluous, and the same number of offspring could be produced by half the population. However, the interaction between individual optimization and dynamics at the population level stabilizes the sex ratio.

The key point of the scenario as developed in this section is that sexual differentiation does not arise in a single step, from parthenogenesis, that is, the production of offspring without a mating partner, to sexual recombination between distinct genders, which would create the puzzle of the two-fold cost of sex, but in two or three steps, first from parthenogenic to recombining, but sexually undifferentiated individuals, and only then through gametic differentiation to sexual differentiation of individuals. None of these steps poses any principal puzzle, and each can be analyzed in terms of optimization.

5.2 Variational Methods

In this section, we shall present the mathematical theory of the optimization of continuous variables, the calculus of variations. We shall only outline the basic ideas, referring to [23, 74, 121], for instance, for more details and many further examples.

The theory of evolution of Darwin [32] and Wallace explains present biological structures as the result of fitness differences in ancestral populations. Therefore, evolutionarily stabilized structures, that is, those not presently undergoing a transition as an adaptation to new external pressures, can be usefully analyzed as maximizing fitness, or more concretely, as optimizing a certain property conducive to fitness. An example of the latter is optimal foraging theory, for instance used for explaining the behavior and traits of human ancestral hunter gatherer populations.

Speciation, that is, the emergence of a new daughter species from an existing one, or the splitting of one ancestral species into two or more new ones, can occur when there are different fitness peaks, that is, trait combinations that represent local fitness maxima. These fitness peaks can be occupied by different species, each of which represents an adaptation to a particular such peak, see [47].

Also in economics, optimization under constraints is a pervasive principle. Microeconomic theory analyzes economic behavior in terms of the maximization of profit or the minimization of costs, or even as the maximization of utility, even though the latter is an empirically and theoretically somewhat problematic concept.

Formally, such problems are posed as variational problems

$$\min_{x} F(x) \quad \text{under the constraints } G_i(x) \leq c_0, \ i = 1, ..., m \qquad (5.2.45)$$

where F is the function to be optimized[1] and G_i, $i = 1, ..., m$, are the functions expressing the constrained quantities. Typically, however, the constraints are a little flexible and not completely rigid, and one may therefore impose them by penalizing large values of the constrained quantities, i.e. formulate the problem

$$\min_x (F(x) + \sum_{i=1}^m \kappa_i G_i(x)) \tag{5.2.46}$$

where the coefficients κ_i, $i = 1, ..., m$, are the relative weights of the constraints and are typically quite large. Formally, the variational problem (5.2.45) is the limit of the problem (5.2.46) for $\kappa_i \to \infty$, $i = 1, ..., m$, but the solutions of (5.2.46) need not always converge to a solution of (5.2.45), for example if different spatial or temporal scales are involved. For the relevant technical issues, we refer for example to [31] or [74].

The mathematical field called calculus of variations is based on the variational principle that optimality in the large implies optimality in the small. The calculus of variations is concerned with problems where the local contributions add up to the global objective function. These are problems of the type

$$F(u) = \int_M f(\xi, u(\xi), Du(\xi))d\xi \to \min \text{ under constraints of the form} G^i(u) = \gamma^i$$
$$\tag{5.2.47}$$

where Du denotes derivative of u and M is the region where the arguments of u vary. For simplicity, we shall assume that M is some open and bounded subset of Euclidean space $\mathbb{R}^d$, although more general cases can be treated by the same type of reasoning. u is a real-valued function on M, or more generally, it takes values in some Euclidean space $\mathbb{R}^n$ or in some more general target, but we shall not consider all possible generalizations here. In the classical case, the constraints are also of integral type, that is

$$G^i(u) = \int_M g^i(\xi, u(\xi), Du(\xi))d\xi \tag{5.2.48}$$

The basic result of Euler says that a necessary condition for a (sufficiently smooth) minimizer is the differential equation

$$\sum_j \frac{d}{d\xi^j}\left(\frac{\partial f}{\partial p_j} + \sum_i \lambda_i \frac{\partial g^i}{\partial p_j}\right) - \left(\frac{\partial f}{\partial u} + \sum_i \lambda_i \frac{\partial g^i}{\partial u}\right) = 0, \tag{5.2.49}$$

---

[1] According to the conventions observed in mathematics and physics, we consider here minimization as opposed to maximization problems. Of course, minimizing F is equivalent to maximizing $-F$, and therefore, this does not at all exclude biological applications where fitness is to be maximized.

where p_j is a dummy variable for $\frac{\partial u}{\partial \xi^j}$, that is, we consider the functions $f(\xi^j, u, p_j)$, $g^i(\xi^j, u, p_j)$.

The λ_i are so-called Lagrange multipliers whose determination is part of the problem. They ensure that a minimizer satisfies the constraints $G^i(u) = \gamma^i$. As explained above, one can replace the rigid constraints by penalizing deviations. The equation (5.2.49) is usually called the Euler-Lagrange equation.

Equation (5.2.49) is an infinitesimal condition that is necessary for u to satisfy the global condition of minimizing the functional F under the constraint of prescribed values for the G^i. For the derivation of (5.2.49), we shall follow [67]. We first consider the case where the G^i all vanish, that is, where we simply have a smooth minimizer $u : M \to \mathbb{R}$ of the functional

$$F(v) = \int_M f(\xi, v(\xi), Dv(\xi))d\xi. \tag{5.2.50}$$

In the sequel, we shall assume that not only f, but also all its derivatives that will occur, as well as u and its derivatives, satisfy suitable integrability conditions so that all integrals are well defined and we shall be able to carry out differentiations under the integral sign (the technical details can be found in [67]). The basic idea is the following. We consider a smooth variation $\phi : M \to \mathbb{R}$ of u. More precisely, since u minimizes F in (5.2.50), we have

$$F(u) \leq F(u + t\phi) \tag{5.2.51}$$

for sufficiently small $|t|$ (that is, we shall only utilize the fact that u is a local minimizer, that is, we only use $F(u) \leq F(v)$ for all v that are sufficiently close to u, w.r.t. some suitable norm that we do not specify here). We also assume that ϕ vanishes near the boundary of M. From this, we derive the infinitesimal condition

$$\frac{d}{dt} F(u + t\phi)_{|t=0} = 0. \tag{5.2.52}$$

In other words, we consider the one-dimensional problem $\Phi(t) = F(u + t\phi)$ as a function of the real variable t, and this then has a minimum at $t = 0$. We now carry out the differentiation in (5.2.52).

$$\begin{aligned}
0 &= \frac{d}{dt} \int_M f(\xi, u(\xi) + t\phi(\xi), Du(\xi) + tD\phi(\xi))d\xi_{|t=0} \\
&= \int_M \Big(\frac{\partial f}{\partial u}(\xi, u(\xi) + t\phi(\xi), Du(\xi) + tD\phi(\xi))\phi(\xi) \\
&\quad + \sum_{j=1}^{d} \frac{\partial f}{\partial p^j}(\xi, u(\xi) + t\phi(\xi), Du(\xi) + tD\phi(\xi))\frac{\partial \phi(\xi)}{\partial \xi^j}\Big)d\xi_{|t=0}
\end{aligned}$$

$$= \int_M (\frac{\partial f}{\partial u}(\xi, u(\xi), Du(\xi)) - \sum_{j=1}^{d} \frac{d}{d\xi^j} \frac{\partial f}{\partial p^j}(\xi, u(\xi), Du(\xi)))\phi(\xi)d\xi \quad (5.2.53)$$

Here, in the last, the crucial, step, we have integrated by parts in order to shift the derivative from ϕ to the integrand. We have been using the assumption that ϕ is smooth in M and that it vanishes near the boundary of M, in order to avoid additional boundary terms. The next step in the reasoning is that (5.2.53) has to hold for *all* such ϕ. From an approximation argument that we skip, one then derives that

$$0 = \frac{\partial f}{\partial u}(\xi, u(\xi), Du(\xi)) - \sum_{j=1}^{d} \frac{d}{d\xi^j} \frac{\partial f}{\partial p^j}(\xi, u(\xi), Du(\xi)). \quad (5.2.54)$$

This is equivalent to (5.2.49) for the case where all the g^i vanish. For the general case, we introduce Lagrange multipliers λ_i and consider

$$\frac{d}{dt}(F(u + t\phi) + \sum_i \lambda_i (G^i(u + t\phi) - \gamma^i))_{|t=0} = 0 \quad (5.2.55)$$

for all the ϕ as before. From this, by the same strategy, we then obtain the general case of (5.2.49). We omit the details and refer to [67] again for the method of Lagrange multipliers.

It should be pointed out that (5.2.49) is only a necessary, but not a sufficient condition for a minimum. In fact, we already know from basic calculus that a first order condition like (5.2.52) only identifies a critical point. Such a critical need not even be a local minimum. For checking whether a critical point is a local minimum, one needs to look at second derivatives. Such a strategy also applies in the calculus of variations, but developing this issue is beyond our present scope.

We now discuss some examples for which a detailed treatment can be found in [74]. We shall only consider the case where $d = 1$ so that the index j is not necessary. In order to simplify the notation, we denote derivatives by subscripts; for instance $u_\xi = \frac{du}{d\xi}$.

1. For a positive function $\gamma : [0, 1] \times \mathbb{R} \to \mathbb{R}$, we consider

$$F(u) = \int_0^1 \frac{\sqrt{1 + u_\xi(\xi)^2}}{\gamma(\xi, u(\xi))}d\xi. \quad (5.2.56)$$

The corresponding Euler-Lagrange equation (5.2.49) then is

$$0 = \frac{d}{d\xi} \frac{u_\xi(\xi)}{\gamma(\xi, u(\xi))\sqrt{1 + (u_\xi(\xi))^2}} + \frac{\gamma_u}{\gamma^2}\sqrt{1 + u_\xi(\xi)^2}$$

$$= \frac{u_{\xi\xi}}{\gamma\sqrt{1+u_\xi(\xi)^2}} - \frac{u_\xi(\xi)^2 u_{\xi\xi}(\xi)}{\gamma(\sqrt{1+u_\xi(\xi)^2})^3} - \frac{\gamma_\xi}{\gamma^2}\frac{u_\xi(\xi)}{\sqrt{1+u_\xi(\xi)^2}}$$

$$- \frac{\gamma_u}{\gamma^2}\frac{u_\xi^2(\xi)}{\sqrt{1+u_\xi(\xi)^2}} + \frac{\gamma_u}{\gamma^2}\sqrt{1+u_\xi(\xi)^2},$$

which yields

$$0 = u_{\xi\xi}(\xi) - \frac{\gamma_\xi}{\gamma}u_\xi(\xi)(1+u_\xi(\xi)^2) + \frac{\gamma_u}{\gamma}(1+u_\xi(\xi)^2). \tag{5.2.57}$$

When γ is constant, the solutions of (5.2.57) are the straight lines $u_{\xi\xi}(\xi) \equiv 0$. There are also other cases where (5.2.57) can be solved explicitly. For example, for $\gamma = \sqrt{2gu(\xi)}$, with g a positive constant, the integrand $f(u(\xi), u_\xi(\xi)) = \frac{\sqrt{1+u_\xi(\xi)^2}}{\sqrt{2gu(\xi)}}$ satisfies

$$\frac{d}{d\xi}(f - u_\xi f_p) = u_\xi(f_u - \frac{d}{d\xi}f_p) = 0 \tag{5.2.58}$$

by (5.2.49). Therefore, every solution satisfies

$$f(u(\xi), u_\xi(\xi)) - u_\xi(\xi)f_p(u(\xi), u_\xi(\xi)) = c \tag{5.2.59}$$

for some constant c which yields

$$u(1 + u_\xi^2) = \frac{1}{2gc^2}. \tag{5.2.60}$$

from which one can determine u.

2. We consider

$$F(u) = \int_{-1}^{1} (1 - u_\xi(\xi)^2)^2 d\xi \tag{5.2.61}$$

with the boundary conditions

$$u(-1) = 1 = u(1). \tag{5.2.62}$$

Obviously, $F(u) \geq 0$. For $u(\xi) = |\xi|$ (which satisfies (5.2.62)), we have $F(u) = 0$. Hence, this u is a minimizer. It is Lipschitz continuous, but not differentiable at 0, and therefore, the Euler-Lagrange equation (5.2.47) is not meaningful in the classical sense. In fact, any Lipschitz function u with the boundary conditions (5.2.62) and

$$|u_\xi(\xi)| = \pm 1 \tag{5.2.63}$$

almost everywhere is a minimizer of F. That is, for a minimizer u, the value of the derivative can switch arbitrarily often between $+1$ and -1. This degeneracy could then be remedied by penalizing such switching, but there is no smooth minimizer, that is, no smooth function u that satisfies the boundary values (5.2.62) and realizes the infimum $F(u) = 0$. Thus, the classical approach of finding a minimizer of F by solving the Euler-Lagrange equation (5.2.49) runs into problems.

3. We consider

$$F(u) = \int_{-1}^{1} u_\xi(\xi)^2 \xi^4 d\xi \tag{5.2.64}$$

with the boundary conditions

$$u(-1) = -1, \quad u(1) = 1. \tag{5.2.65}$$

For

$$u_n(\xi) = \begin{cases} -1 & \text{for} \quad 1 \le \xi < -\frac{1}{n} \\ n\xi & \text{for} \ -\frac{1}{n} \le \xi \le \frac{1}{n} \\ 1 & \text{for} \ \frac{1}{n} < \xi < 1 \end{cases} \tag{5.2.66}$$

we have $\lim_{n\to\infty} F(u_n) = 0$, but for every function u satisfying (5.2.65), we have $F(u) > 0$. Thus, the infimum of the functional F of (5.2.64) with the boundary conditions (5.2.65) is not achieved. Thus, this variational problem has no solution.

4. We now consider an example with a penalization. Let W be a nonnegative function, with $W(\xi) = 0$ precisely for the two values $\xi = a$ and $\xi = b$, with $0 < a < b < 1$. We look at

$$F(u) = \int_{0}^{1} W(u(\xi))d\xi \tag{5.2.67}$$

with the constraint

$$\int_{0}^{1} u(\xi)d\xi = c, \tag{5.2.68}$$

for some given $c > 0$. A minimizer u then satisfies

$$u_\gamma = \begin{cases} a & \text{in } A_1 \subset [0, 1] \\ b & \text{in } A_2 \subset [0, 1] \end{cases} \tag{5.2.69}$$

with two disjoint sets A_1, A_2, whose union is $[0, 1]$ and which satisfy

$$a \, \text{meas}(A_1) + b \, \text{meas}(A_2) = c. \tag{5.2.70}$$

The problem is that the sets A_1 and A_2 are otherwise quite arbitrary, that is, u can switch arbitrarily between the values a and b. To remedy this, we consider the functionals

$$F_n(u) = \int_0^1 \left(nW(u(\xi)) + \frac{1}{n}|u_\xi(\xi)|^2 \right) d\xi \qquad (5.2.71)$$

They can be minimized, and for $n \to \infty$, the solutions u_n converge to minimizers of the functional

$$F_0(u) = 2\gamma(\text{number of points of discontinuity of } u) \qquad (5.2.72)$$

when u is a piecewise constant function; here $\gamma = \int_a^b W^{1/2}(s)ds$ is an irrelevant normalizing constant. Thus, even though the penalty disappears in the limit $n \to \infty$, its effect is preserved in the limits of minimizers. Thus, its regularizing effect persists.

5. We finally discuss the famous isoperimetric problem (see e.g. [25]). In its simplest form, it requires to maximize the area A of a bounded region Ω in the plane whose boundary has a prescribed length L. We have

$$A = \int_\Omega du\,dv \qquad (5.2.73)$$

$$L = \int_{\partial\Omega} ds \qquad (5.2.74)$$

where u, v are coordinates in the plane and ds is the line element of the boundary of Ω. With a Lagrange multiplier λ, we therefore consider

$$-\int_\Omega du\,dv + \lambda \int_{\partial\Omega} ds. \qquad (5.2.75)$$

Assuming that everything is smooth, we parametrize the boundary $\partial\Omega$ as

$$s(\xi) = (u(\xi), v(\xi)) \qquad (5.2.76)$$

with smooth functions u, $v : [0, 1] \to \mathbb{R}$. These functions need to close up, that is, $u(0) = u(1)$ and $v(0) = v(1)$ because the boundary of Ω is a closed curve. The exterior unit normal vector of $\partial\Omega$ is then given by

$$n := \left(\frac{\dot{v}(\xi)}{\sqrt{\dot{u}(\xi)^2 + \dot{v}(\xi)^2}}, -\frac{\dot{u}(\xi)}{\sqrt{\dot{u}(\xi)^2 + \dot{v}(\xi)^2}} \right) \qquad (5.2.77)$$

We recall the theorem of Gauss and Stokes,

$$\int_\Omega \operatorname{div} X(u, v) du dv = \int_{\partial\Omega} X \cdot n ds \qquad (5.2.78)$$

for a vector field $X = (X^1, X^2)$ with $\operatorname{div} X(u, v) := \frac{\partial X^1}{\partial u} + \frac{\partial X^2}{\partial v}$ and the scalar product $X \cdot n := X^1 n^1 + X^2 n^2$. We can then use the vector field $X = (u, v)$ with $\operatorname{div} X = 2$ to convert the area into a boundary integral, observing that by (5.2.76), $ds = \sqrt{\dot{u}(\xi)^2 + \dot{v}(\xi)^2} d\xi$,

$$A = \int_0^1 (u(\xi)\frac{\dot{v}(\xi)}{2\sqrt{\dot{u}(\xi)^2 + \dot{v}(\xi)^2}} - v(\xi)\frac{\dot{u}(\xi)}{2\sqrt{\dot{u}(\xi)^2 + \dot{v}(\xi)^2}})\sqrt{\dot{u}(\xi)^2 + \dot{v}(\xi)^2} d\xi.$$

(5.2.75) then becomes

$$I(u, v, \lambda) := \int_0^1 (-\frac{u(\xi)\dot{v}(\xi)}{2} + \frac{v(\xi)\dot{u}(\xi)}{2}) d\xi + \lambda \int_0^1 \sqrt{\dot{u}(\xi)^2 + \dot{v}(\xi)^2} d\xi.$$

$$(5.2.79)$$

Introducing the dummy variables p for $\dot{u}$ and q for $\dot{v}$, we write this as

$$I(u, v, \lambda) = \int_0^1 (-\frac{uq}{2} + \frac{vp}{2}) d\xi + \lambda \int_0^1 \sqrt{p^2 + q^2} d\xi \qquad (5.2.80)$$

$$= \int_0^1 (f(u, v, p, q) + g(u, v, p, q)) d\xi.$$

The Euler-Lagrange equations for (5.2.80) are

$$\frac{d}{d\xi}(f_p + \lambda g_p) = f_u + \lambda g_u$$

$$\frac{d}{d\xi}(f_q + \lambda g_q) = f_v + \lambda g_v$$

$$p^2 + q^2 = L^2,$$

that is,

$$q + \lambda\frac{d}{d\xi}\frac{p}{\sqrt{p^2 + q^2}} = 0$$

$$p - \lambda\frac{d}{d\xi}\frac{q}{\sqrt{p^2 + q^2}} = 0$$

$$p^2 + q^2 = L^2.$$

The solution is

$$\lambda = \frac{L}{4\pi}$$
$$p(\xi) = L \cos 2\pi\xi$$
$$q(\xi) = L \sin 2\pi\xi,$$

that is,

$$u(\xi) = \frac{L}{2\pi} \sin 2\pi\xi, \quad v(\xi) = -\frac{L}{2\pi} \cos 2\pi\xi, \tag{5.2.81}$$

that is, a circle. We do not verify here that this provides the absolute maximum of our problem, but rather refer to the literature, e.g. [25]. The area then is $A = \frac{L^2}{4\pi}$, that is, it grows quadratically as a function of the length L. Instead of maximizing the area for a given boundary length, one can also minimize the boundary length for a given area. This leads to an equivalent problem, as the reader should convince herself.

Analogously, the solid body that maximizes the volume V with a given surface area A is the solid sphere (round ball). Here, the relation between volume and surface area is

$$V \sim A^{3/2}. \tag{5.2.82}$$

As before, one can equivalently minimize the surface area for a given volume. For instance, if for a biological organism, the temperature loss to the environment is proportional to its surface, then, for a given volume, it should try to adopt a round shape, of course, to the extent that this is compatible with its other requirements. Also, the scaling law (5.2.82) tells us that such an optimization becomes better, in the sense that it requires relatively less surface in relation to the volume, when the organism gets bigger, that is, increases its volume. Such scaling laws can therefore be helpful for understanding the sizes that organisms develop. Again, of course, while according to the isoperimetric relationship (5.2.82), bigger is better, this must be related to metabolic, structural and other requirments that affect the fitness of an organism.

In fact, biological scaling laws seem to be quite general, but they are different from the simple relationship (5.2.82), because they rather depend on the properties of metabolic flows, see [21].

Exercises for This Chapter

1. It has been observed that in human populations male babies on average slightly outnumber female ones, that is, at birth there is a sex ratio of 102:100 or so. Evolutionary biology suggests that the reason for this lies in the slightly higher

mortality rate of male infants. Develop this argument within the formal framework provided in this chapter. In particular, find out the equilibrium sex ratio for a given infant mortality rate ratio.

2. Find a variational problem whose Euler-Lagrange equation is the Laplace equation $\Delta u = 0$.

3. Formulate a variational problem that can identify the most energy efficient route between two points for migrating birds when travelling distances and wind forces along the route are taken into consideration.

Chapter 6
Population Genetics

Abstract
Questions:

- How does the distribution of alleles (genetic variants) change over time in a population when those alleles are randomly passed on to offspring?

This last chapter draws upon all the different methods discussed in the preceding, discrete structures, stochastics, analysis, and geometry. It introduces mathematical population genetics, the theory of the time course of the distribution of alleles in a population in the presence of mutation, selection, and recombination. The basic Wright-Fisher model is a discrete stochastic processes. In order to understand it better, it is advantageous to pass to its diffusion approximation which leads to a partial differential equation. For understanding this differential equation in turn a geometric approach is insightful.

Population genetics is concerned with the stochastic dynamics of allele frequencies in a population. Alleles are alternative values at genetic loci. Thus, at one such locus, there are finitely many possible values that we can label by $1, \ldots, k$, the simplest nontrivial case being $k = 2$ which usually already shows all the features of interest. Different individuals in the population may have different values, and the relative frequency of the value i (at some given time) is denoted by p_i. Thus, p is a probability measure on $S_k := \{1, \ldots, k\}$, that is,

$$\sum_{i=1}^{k} p_i = 1. \tag{6.0.1}$$

The population is evolving in time, and members pass on genes to their offspring, and the allele frequencies p_i then change in time because of the mechanisms of selection, mutation and recombination. In the simplest case, one has a population with nonoverlapping generations. That means that we have a discrete time index n, and for the transition from n to $n + 1$, the population P_n produces a new population P_{n+1}. More precisely, members of P_n can give birth to offspring that inherit their alleles. This process involves potential sources of randomness. Mutation means that

J. Jost, *Mathematical Methods in Biology and Neurobiology*, Universitext, DOI: 10.1007/978-1-4471-6353-4_6, © Springer-Verlag London 2014

an allele may change to another value in the transition from parent to offspring. Selection means that the chances of producing offspring vary depending on the value of the allele in question, as some alleles may be fitter than others. Recombination takes place in sexual reproduction, that is, when each member of the population has two parents. It is then determined by chance which allele value she inherits when the two parents possess different alleles at the locus in question.

Since the preceding are stochastic effects, the future frequencies become probabilities, that is, instead of saying that $p_i N$ of the N individuals in the population carry the allele i, we rather need to say that the probability of finding the allele i at the locus in question is p_i. A key point of the mathematical framework for population genetics (and for many other fields) then is the assumption that these probabilities (while expressing stochastic effects) change in time according to deterministic rules. As already indicated, one typically considers a finite population with a discrete time dynamics. It is often useful, however, to pass to the limit of an infinite population. In order to compensate for the growing size, one then needs to make the time steps shorter and pass to continuous time.

We start with the simplest situation where we consider only one locus; essentially, this means that we assume that the dynamics at this locus are independent of what happens at other loci. This is, of course, unrealistic, but it leads to the simplest models. We can then try to generalize these models in subsequent steps.

6.1 Mutation, Selection and Recombination

The models that we are going to discuss all make a number of simplifying assumptions of varying biological plausibility, in order to make a formal treatment possible.

We consider a population P_n that is changing in discrete time n with nonoverlapping generations, that is, the population P_{n+1} consists of the offspring of the members of P_n. In particular, we neglect the issue of migration here. Each individual in the population is represented by its genotype x. We assume that the genetic loci of the different members of the population are in one-to-one correspondance with each other. Thus, we have loci $\alpha = 1, \dots A$. In the haploid case, at each locus, there can be one of k_α possible alleles. Thus, a genotype is of the form $x = (x^1, \dots x^A)$, where $x^\alpha \in \{1, 2, \dots, k_\alpha\}$. In the diploid case, at each locus, there are two alleles, which could be the same or different. We are interested in the distribution of genotypes x in the population and how that distribution changes over time through the effects of mutation, selection, and recombination.

The baseline situation might be that each member of P_n by itself, that is, without recombination, produces one offspring that is identical to itself. In that case, nothing changes in time. This baseline situation can then be varied in three respects:

1. The offspring is not necessarily identical to the parent (mutation).
2. The number of offspring an individual produces or may be expected to produce varies with that individual's genotype (selection).
3. Each individual has two parents, and its genotype is assembled from the genotypes of its parents (sexual recombination).

In fact, item 3 has two aspects:

a. Each allele is taken from one of the parents in the haploid case. In the diploid case, each parent produces gametes, which means that she chooses one of her two alleles at the locus in question and gives it to the offspring. Of course, this choice is made for each offspring, so that different descendents can carry different alleles.
b. Since each individual has many loci that are linearly arranged on chromosomes, alleles at neighboring loci are not passed on independently.

Thus, we want to model how the three mechanisms of mutation, selection and recombination change the distribution of genotypes in the population over time.

Again, in the basic models, these three mechanisms are assumed to be independent of each other. For instance, it is assumed that mutation rates do not work in favor of selectively superior alleles.

The mechanism of selection leads to the issue of fitness. This is a difficult concept; in particular, it is not clear what the unit of fitness is, whether it is the allele or the genotype or the ancestor of a lineage, or in groups of interacting individuals even some higher order unit. Some of these issues have been discussed in [66]. Also, when discussing the concept of fitness, one needs to distinguish between the actual number of offspring produced by a member of a past population and the expected number of offspring of a member of the present or a future population. Moreover, in the latter case, one needs to specify on what those expectations are conditioned. This issue seems to be sometimes overlooked in the biological literature.

If we employ the term "fitness" loosely for the moment and speak of the fitness of an individual as the (expected) number of offspring it produces, this fitness may depend not only on that individual's own genotype, but perhaps also on the distribution of genotypes in the population. For instance, as we have seen in Sect. 5.1, if we have a dioecious population, that is, one with different sexes so that a female has to pair with a male individual to produce offspring, then the individuals of the minority sex have a higher (expected) fitness than the majority representatives.

In any case, the fitness will depend on the environment that the population is situated in. In the simplest case, this environment is uniform (that is, the same for everybody), static (not varying in time) and independent of the population. Obviously, this is a gross simplification. In order to dispense with the stationarity assumption, one may consider random environments, that is, ones that are stochastically fluctuating in time. In the simplest such case again, one then assumes that the environments randomly varies according to some static distribution of possible environments. In other cases, one may assume a systematic trend like global warming. Also, often the crucial aspect of the environment that the fitness will depend on is access to resources. If those resources are limited, then the fitness will ultimately decrease with growing population size (this effect has been included in some of the dynamical population models treated in Sect. 4.3.1, for instance the Fisher equation (4.3.62)). A natural concept in that situation is the carrying capacity of the environment. It becomes more interesting when we consider two different populations that interact with each other, thereby each contributing to the other's environment. For each population, we then have a dynamic interaction between that population and its environment (from a different perspective, this issue was addressed by the Lotka-Volterra equations in Sect. 4.3.1). Finally, in order to dispense with the uniformity assumption for the environment, we may add dimensions along which the environment varies. These dimensions may correspond to physical space, but may also reflect other differences between the situation and the behavior of individuals.

Also, in reality, the relation between an individual's genotype and its fitness is rather indirect. Through a developmental process, from a genotype, and perhaps with substantial contributions from epigenetic inheritance, in intricate interaction with the environment, an adult phenotype is produced, and that phenotype then, together with environmental conditions, determines the fitness of an individual.

In view of all these additional layers of complexity, the situation for mathematically tractable formal models may look hopeless, but in fact, this is not quite so. Even the simplest models do offer some important insights, and many of the complications just described can be successfully included into formal models. Of course, the development of models should go hand in hand with the clarification of concepts. The most basic ones are the concepts of the gene and of the species, together with a clarification of the issue and the unit of fitness.

Thus, to start with our models, a population is considered as a distribution of genotypes. Since we do not yet know the composition of future populations, we need to work with probability distributions. Thus, $p_n(x)$ is the probability that an individual in generation n carries the genotype x. We are then interested in the dynamics of that probability distribution p_n in time n.

In order to model mutations, we assume that we have an operator represented by a matrix $M = (m_{xy})$ where x, y range over the possible genotypes and m_{xy} is the probability that genotype y mutates to genotype x. This mutation probability should depend on in how many loci x and y carry different alleles. Let $d(x, y)$ be that number. In the simplest case, one assumes that there is a uniform rate m with which a mutation takes place at each locus, and this is the same for all the alleles that may be present at that locus. That is, an allele i at the locus α mutates with probability m. It then turns with probability $\frac{m}{k_\alpha - 1}$ into the allele $j \neq i$. If we want to simplify this, we assume that $k_\alpha \equiv k$ is independent of α, that is, the number of possible alleles is the same at each locus α. If we want to simplify it still further, we assume $k = 2$. We then have

$$m_{xy} = (\frac{m}{k-1})^{d(x,y)}(1 - \frac{m}{k-1})^{A-d(x,y)} \tag{6.1.2}$$

and

$$m_{xy} = m^{d(x,y)}(1 - m)^{A-d(x,y)} \tag{6.1.3}$$

in case $k = 2$.

In contrast to mutation, recombination is a binary operation, that is, an operation that takes two parent genotypes y, z as arguments to produce one offspring genotype x. We consider the case of monoecious individuals with haploid genotypes for the moment. "Monoecious" means that we do not have separate sexes, that is, any individual can pair with each other one to produce offspring. "Haploid" means that each individual carries one allele at each locus. An offspring is then formed through recombination by choosing at each locus the allele that one of the parents carries there. When the two parents carry different alleles at the locus in question, we have to decide by a selection rule which one to choose. This selection rule is represented by a mask μ, a binary string of length A. A 1 at position α means that the allele is taken from the first parent, say y, and a 0 signifies that the allele is taken from the second parent, say z. For instance, for $k = 2$ so that we can also represent each genotype as a binary string of length A, and for $A = 5$, the mask 10010 produces from the parents $y = 11001$ and $z = 01010$ the offspring $x = 11000$. We can then write the recombination operator as

$$R_{xyz} = \sum_{\mu} p_r(\mu) C_{xyz}(\mu), \tag{6.1.4}$$

where $C_{xyz}(\mu)$ is the recombination scheme for the mask μ and $p_r(\mu)$ is the probability for the mask μ. In the simplest case (still assuming $k = 2$), all the possible 2^A masks are equally probable, and consequently, at each locus, the offspring chooses an allele from either parent with probability $1/2$, independently of the choices at the other loci. Another situation occurs in the so-called cross-over models where the only possible masks are of the form $\mu_c = 11\ldots 100\ldots 0$, that is, at the first $a(\mu_c)$, the allele from the first parent is chosen, and at the remaining $A - a(\mu_c)$, that from the second parent. We then have $A + 1$ possible such masks μ_c, and we may wish to assume again that each of those is equally probable. Such crossover models are employed in genetic algorithms, a biologically inspired population based optimization scheme, see e.g. [51, 91]. The key point is that one wishes to consider fitness functions where the fitness is not simply an additive contribution from the allele values at the individual loci, but where it may depend on certain allele combinations. For instance, it may be best for an individual to carry either 101 or 010 at the first three loci. When an individual with 101 is paired with an individual of type 010, only the masks 111 and 000 lead to an offspring with one of the two favorable allele combinations. If we allow only crossover masks, the probability of that event is significantly enhanced. Thus, with such a cross-over scheme, there is much better chance that favorable allele combinations are preserved in the population.

The preceding consideration assumed random mating, that is, individuals get paired at random to produce offspring. As an alternative, one may also wish to assume assortative mating, that is, the chances for two individuals to form such a pair are higher the more genetically similar they are. Such a mechanism will then lead to some clustering of the genotypes in the population.

In the diploid case, each individual carries two alleles at each locus, one from each parent. When offspring is produced, it is then randomly decided which of the two alleles of each parent is carried over. Otherwise, the scheme is the same as in the haploid case.

With recombination alone, some alleles may disappear from the populations, and in fact, with probability 1, in the long term, only one allele will survive at each site, due to random genetic drift, that is, because the parents that produce offspring are randomly selected from the population. This is because it may happen that no carrier of a particular allele is chosen at a given time or that none of the chosen recombination masks preserves that allele when the mating partner carries a different allele at the locus under consideration. That would then lead to the ultimate extinction of that allele. However, when we also allow for mutations, an allele that is not present in the population at time n may reappear at time $n + 1$.

The order in which the mutation and recombination operators are applied is not important in most models.

So far, individuals in generation n were randomly selected as parents of individuals in generation $n + 1$. Formally, for each individual in generation $n + 1$, we sample the generation n to choose its one or two parents. Again, the simplest case is to take sampling with replacements, that is, each individual in generation n may in principle produce arbitrary many offspring. The question then emerges by what rule the population at time n is sampled. This is the effect of selection. In the simplest case, selection means that the probability of an individual in generation n to be chosen as a parent is some function of its genotype. In other words, there is some probability distribution $p_s(x)$ on the space of genotypes x. Here, we make the simplifying assumption that this probability depends only on the individual genotype, but not on the composition of the population. Also, we assume a static environment, that is, $p_s(x)$ does not depend on n. All these assumptions can be abandoned, but we wish to start with the simplest scenario.

Another issue is whether we keep the population size constant, say $\equiv N$, or let it vary. In the latter case, the population might become extinct in finite time. The basic model here is the Galton-Watson branching process discussed in Sect. 3.4, where each individual produces some number m of offspring with probability $p(m)$. We have, of course, $\sum_m p(m) = 1$, and the expected number of offspring is $E(m) = \sum_m mp(m)$. Obviously, a necessary (but in general not sufficient) condition for the population to survive is $E(m) \geq 1$. Again, extensions of this model along the above lines are possible and meaningful. For instance, there may be different types of individuals with different survival and reproduction chances, and this may depend on the composition of the population, like the ratio between male and female individuals.

6.2 The Wright-Fisher Model and its Diffusion Approximation

In population genetics, as discussed in the previous section, one considers the effects of recombination, selection, mutation, and perhaps others like migration on the distribution of alleles in a population. References are [43, 102].

We start with the Wright-Fisher model for the effects of genetic drift and consider a diploid population of size N. At a given locus, there could be either one of two alleles A_1, A_2. Thus, an individual can be a homozygote of type $A_1 A_1$ or $A_2 A_2$ or a heterozygote of type $A_1 A_2$ or $A_2 A_1$—but we consider the latter two as the same—at the locus in question. The population reproduces in discrete time steps, and each individual in generation $t + 1$ inherits one allele from each of its parents. When a parent is a heterozygote, each allele is chosen with probability $1/2$. Here, for each individual in generation $t + 1$, randomly two parents in generation t are chosen, as in Sect. 2.3.3. Thus, the alleles in generation $t + 1$ are chosen by random sampling with replacement from those in generation t. The quantity of interest is the number $X(t)$ of alleles A_1 in the population at time t. This number then varies between 0 and $2N$. The transition probability then is given by the binomial distribution,

$$p(X(t+1) = j | X(t) = i) = \binom{2N}{j} (\frac{i}{2N})^j (1 - \frac{i}{2N})^{2N-j} \text{ for } i, j = 0, \ldots, 2N.$$

$$(6.2.1)$$

Whenever $X(t)$ takes the value 0 or $2N$, it will stay there for all future times. Eventually, this will happen almost surely.

This is the basic model. One can then derive expressions for the expected time for the allele A_1 to become either fixed, that is, $X(t) = 2N$ or become extinct, $X(t) = 0$, given its initial number $X(0)$.

We now include selection. We thus assume selective differences between the two alleles. More precisely, let the fitness values of

$$\begin{array}{ccc} A_1 A_1 & A_1 A_2 & A_2 A_2 \\ \text{be } 1 + s & 1 + sh & 1 \end{array} \qquad (6.2.2)$$

Here, $0 \leq h \leq 1$, the simplest case being $h = 1/2$. Then (6.2.1) gets replaced by

$$p(X(t+1) = j | X(t) = i) = \binom{2N}{j} (\eta_i)^j (1 - \eta_i)^{2N-j} \text{ for } i, j = 0, \ldots, 2N$$

$$(6.2.3)$$

with

$$\eta_i = \frac{(1+s)i^2 + (1+sh)i(2N-i)}{(1+s)i^2 + 2(1+sh)i(2N-i) + (2N-i)^2}. \qquad (6.2.4)$$

When instead mutation occurs, but no selection, we still get other coefficients. Let us assume that A_1 mutates to A_2 with rate u, A_2 to A_1 with rate v. We then have

$$p(X(t+1) = j | X(t) = i) = \binom{2N}{j} (\psi_i)^j (1 - \psi_i)^{2N-j} \text{ for } i, j = 0, \ldots, 2N$$

$$(6.2.5)$$

with

$$\psi_i = \frac{i(1-u) + (2N-i)v}{2N}. \qquad (6.2.6)$$

For the diffusion approximation, we assume that

$$\alpha := 2Ns, \quad \beta_1 := 2Nu, \quad \beta_2 := 2Nv \tag{6.2.7}$$

are all of order 1, and we put

$$\delta t := \frac{1}{2N}, \quad x := \frac{i}{2N}, \quad x + \delta x := \frac{j}{2N}. \tag{6.2.8}$$

From the above model, we then obtain

$$E(\delta x | x) = \frac{\alpha x(1-x)(x + h(1-2x)) - \beta_1 x + \beta_2(1-x)}{2N} + o(\frac{1}{N})$$

$$=: \frac{f(x)}{2N} + o(\frac{1}{N}) \tag{6.2.9}$$

$$\mathrm{var}(\delta x | x) = \frac{x(1-x)}{2N} + o(\frac{1}{N}) \tag{6.2.10}$$

$$E(\delta x^3) = o(\frac{1}{N}). \tag{6.2.11}$$

For $N \to \infty$, we then obtain for the density $h(x, t)$

$$\frac{\partial}{\partial t} h(x, t) = \frac{1}{2} \Delta(x(1-x)h(x, t)) - \frac{\partial}{\partial x}(f(x)h(x, t)). \tag{6.2.12}$$

This has a different diffusion term than (4.5.23). We do not discuss here the detailed analytical derivation of the diffusion approximation Eq. (6.2.12), which is also called a Fokker-Planck or Kolmogorov forward equation. We refer to [22, 43] and the forthcoming detailed treatment in [62]. The general mathematical theory of diffusion approximations of Markov processes can be found in [41]. In the next section, we rather present the—perhaps somewhat surprising—geometric interpretation of (6.2.12) from which one can understand many aspects of its solutions and therefore also of the solutions of the Wright-Fisher model.

6.3 The Geometry of Probability Distributions

Let $S_k := \{1, \ldots, k\}$ be the finite set of k elements. A measure π on S_k assigns to every $j \in S_k$ a nonnegative number π_j. We also require that at least some π_j are positive. The space of measures on S_k is denoted by $\mathfrak{M}$ (dropping the subscript k for simplicity of notation). For $A \subset S_k$, we put $\pi(A) := \sum_{j \in A} \pi_j$. A probability measure on S_k is a measure p with

$$\sum_j p_j = 1. \tag{6.3.1}$$

We are mostly interested in probability measures, that is, measures that satisfy the normalization (6.3.1). We could then consider the space of probability measures as a subspace of the space of all measures. There is, however, a better way to conceptualize it. The normalization (6.3.1) can be achieved by rescaling a given measure π, that is, by multiplying it by some appropriate factor $\lambda > 0$, namely by $\lambda = (\sum_j \pi_j)^{-1}$. The freedom of rescaling a measure now expresses that we are not interested in absolutes "sizes" $\pi(A)$ of subsets of S_k, but rather only in relative ones, like $\frac{\pi(A)}{\pi(S_k)}$ or $\frac{\pi(A_1)}{\pi(A_2)}$, that is, in relative frequencies. Therefore, we identify the space $\mathfrak{P}$ of probability measures on S_k as the projective space

$$\mathbb{P}^1\mathfrak{M},$$

i.e., the space of all equivalence classes in $\mathfrak{M}$ under multiplication by positive real numbers. Of course, elements of $\mathfrak{P}$ can be considered as measures satisfying (6.3.1), but more appropriately as equivalence of measures giving the same relative sizes of subsets of S_k.

The probability measures on S_k are given by

$$\Sigma^{k-1} := \{(p_1, ..., p_k) : p_j \geq 0 \ \text{ for } j = 1, ..., k, \text{ and } \sum_{j=1}^{k} p_j = 1\}.$$

These form a $(k-1)$-dimensional simplex in the positive cone $\mathbb{R}^k_+$ of $\mathbb{R}^k$. The projective space

$$\mathbb{P}^1\mathbb{R}^k_+,$$

however, naturally is identified with the corresponding spherical sector

$$S^{k-1}_+ := \{(z_1, ..., z_k) : z_j \geq 0 \text{ for } j = 1, ..., k, \ \sum_{j=1}^{k} z_j^2 = 1\}.$$

There is a natural bijection

$$\Sigma^{k-1} \ \rightarrow \ S^{k-1}_+$$
$$(p_1, ..., p_k) \rightarrow (\sqrt{p_1}, ..., \sqrt{p_k}).$$

Thus, if p is a probability measure on S_k, then its square root $\sqrt{p}$ is an element of the unit sphere in $\mathbb{R}^k$.

Definition 6.3.1. Let $p(s)$ be a family of probability distributions depending on the parameters $s = (s_1, \ldots, s_n)$. The Fisher information metric of this family at $s = 0$ is given by

$$E_p(\frac{\partial}{\partial s_\mu} \log p(s) \frac{\partial}{\partial s_\nu} \log p(s)) = \sum_j \frac{\partial}{\partial s_\mu} \log p_j(s) \frac{\partial}{\partial s_\nu} \log p_j(s) p_j \qquad (6.3.2)$$

where E_p denotes the expectation with respect to the probability measure $p = p(0)$ and all derivatives are taken at $s = 0$.

We have

Lemma 6.3.1. *The Fisher metric can be expressed as*

$$E_p(\frac{\partial}{\partial s_\mu} \log p(s) \frac{\partial}{\partial s_\nu} \log p(s)) = -E_p(\frac{\partial^2}{\partial s_\mu \partial s_\nu} \log p(s)), \qquad (6.3.3)$$

again with all derivatives taken at $s = 0$.

Proof. We have

$$\sum_j \frac{\partial}{\partial s_\mu} \log p_j(s) \, p_j(s) = \frac{\partial}{\partial s_\mu} \sum_j p_j(s) = \frac{\partial}{\partial s_\mu} 1 = 0, \qquad (6.3.4)$$

since all the $p(s)$ are probability measures. This implies, when we take all derivatives at $s = 0$ and put $p_j = p_j(0)$,

$$\begin{aligned}
0 &= \frac{\partial}{\partial s_\nu} \left(\sum \frac{\partial}{\partial s_\mu} \log p_j(s) \, p_j(s) \right) \\
&= \sum \frac{\partial^2}{\partial s_\mu \partial s_\nu} \log p_j(s) \, p_j(0) + \sum \frac{\partial}{\partial s_\mu} \log p_j(s) \frac{\partial}{\partial s_\nu} p_j(s) \\
&= \sum \frac{\partial^2}{\partial s_\mu \partial s_\nu} \log p_j(s) \, p_j + \sum \frac{\partial}{\partial s_\mu} \log p_j(s) \frac{\partial}{\partial s_\nu} \log p_j(s) \, p_j.
\end{aligned}$$

Since with $p = p(0)$

$$E_p(\frac{\partial^2}{\partial s_\mu \partial s_\nu} \log p(s)) = \sum_j \frac{\partial^2}{\partial s_\mu \partial s_\nu} \log p_j(s) p_j,$$

Equation (6.3.3) follows. $\square$

Lemma 6.3.2. *We can also write the Fisher metric (6.3.2) as*

$$E_p\left(\frac{\partial}{\partial s_\mu}\log p(s)\frac{\partial}{\partial s_\nu}\log p(s)\right) = \sum_{i=1}^{n}\frac{1}{p_i}\frac{\partial}{\partial s_\mu}p_i\frac{\partial}{\partial s_\nu}p_i, \qquad (6.3.5)$$

where again all derivatives have to be taken at $s = 0$. In particular, the Fisher metric yields a quadratic form on the variations $\frac{\partial}{\partial s_\mu}p(s)|_{s=0}$.

The *proof* follows directly from the definition (6.3.2).

Thus, the metric tensor in the coordinates $p_1,\ldots,p_k$ becomes

$$\begin{pmatrix} \frac{1}{p_1} & 0 & \cdots & 0 \\ 0 & \frac{1}{p_2} & \cdots & 0 \\ \vdots & & & \\ 0 & 0 & \cdots & \frac{1}{p_k} \end{pmatrix}. \qquad (6.3.6)$$

Remark: This metric is called the Shashahani metric in mathematical biology, see [61].

This is simply the metric obtained on the simplex Σ^{k-1} when identifying it with the spherical sector S_+^{k-1} via the map $p = q^2$, $q \in S_+^{k-1}$. If the second derivatives $\frac{\partial^2}{\partial s_\mu \partial s_\nu}p$ vanish, i.e., if $p(s)$ is linear in s, then

$$\sum_{j=1}^{k}\frac{1}{p_j}\frac{\partial}{\partial s_\mu}p_j\frac{\partial}{\partial s_\nu}p_j = \frac{\partial^2}{\partial s_\mu \partial s_\nu}\sum_{j=1}^{k}p_j\log p_j.$$

An interpretation of this is that the negative of the entropy of probability measures is a potential for the metric.

We can also apply the formal tools of Riemannian geometry (see [70]) to the Fisher metric. (6.3.5), (6.3.6), however, is not yet the expression for a Riemannian metric because we have k coordinates $p_1,\ldots,p_k$ on a $(k-1)$-dimensional space. This can be easily corrected, however, by expressing

$$p_k = 1 - \sum_{j=1}^{k-1}p_j. \qquad (6.3.7)$$

By the transformation behavior for a Riemannian metric,

$$g_{ij}(x) = \sum_{\alpha,\beta}\gamma_{\alpha\beta}(y)\frac{\partial y^\alpha}{\partial x^i}\frac{\partial y^\beta}{\partial x^j}, \qquad (6.3.8)$$

when transforming between the coordinates x and $y = y(x)$ (here we take $x = (p_1, \ldots, p_{k-1})$, $y = (p_1, \ldots, p_k)$), and using

$$\frac{\partial p_k}{\partial p_j} = -1 \text{ for } j = 1, \ldots, k - 1, \tag{6.3.9}$$

we obtain the metric tensor g_{ij} in the coordinates $p_1, \ldots p_{k-1}$ as

$$\begin{pmatrix} \frac{1}{p_1} + \frac{1}{p_k} & \frac{1}{p_k} & \cdots & \frac{1}{p_k} \\ \frac{1}{p_k} & \frac{1}{p_2} + \frac{1}{p_k} & \cdots & \frac{1}{p_k} \\ \vdots & & & \\ \frac{1}{p_k} & \frac{1}{p_k} & \cdots & \frac{1}{p_{k-1}} + \frac{1}{p_k} \end{pmatrix}, \tag{6.3.10}$$

with p_k given by (6.3.7). For later purposes, we also need the inverse metric tensor g^{ij} which then becomes

$$\begin{pmatrix} p_1(1 - p_1) & -p_1 p_2 & \cdots & -p_1 p_{k-1} \\ -p_1 p_2 & p_2(1 - p_2) & \cdots & -p_2 p_{k-1} \\ \vdots & & & \\ -p_1 p_{k-1} & -p_2 p_{k-1} & \cdots & p_{k-1}(1 - p_{k-1}) \end{pmatrix}. \tag{6.3.11}$$

6.4 Population Dynamics

We now look at dynamics. Some references for the sequel are [22, 61]. We assume that we have a population consisting of y^i individuals of type i, $i = 1, \ldots, k$. Thus, the relative frequency of type i is

$$p^i := \frac{y^i}{\sum_j y^j}. \tag{6.4.1}$$

Each type then may have some fitness $f_i(y)$ which depends also on the presence and magnitude of other types in the population. This fitness is supposed to express the growth rate of type i.[1] It will thus change in time according to

$$\dot{y}^i = f_i(y) y^i, \text{ for } i = 1, \ldots, k. \tag{6.4.2}$$

When we want to separate what is specific for the type i and what holds uniformly for the entire population, we may also redefine the f_i and write

$$\dot{y}^i = (f_i(y) + f_0(y)) y^i. \tag{6.4.3}$$

---

[1] The concept of fitness is somewhat subtle, but this is not our concern here.

The average fitness of the population then is

$$\bar{f} := \sum_{i=1}^{k} f_i \, p^i . \tag{6.4.4}$$

If we also wanted to account for the uniform growth term f_0, we should put instead

$$\bar{f}_0 := \sum_{i=0}^{k} f_i \, p^i = \bar{f} + f_0 . \tag{6.4.5}$$

We now assume that the f_i, $i = 0, 1, \ldots, k$ depend only on the relative frequencies p^i, but not on the absolute magnitudes y^i. We change our notation and write $f_i(p)$ in place of $f_i(y)$. We can then obtain differential equations for the p^i, the *replicator equations*,

$$\begin{aligned}
\dot{p}^i &= p^i (f_i + f_0 - \sum_{j=1}^{k} p^j (f_j + f_0)) \\
&= p^i (f_i - \sum_{j=1}^{k} p^j f_j) \\
&= p^i (f_i - \bar{f}),
\end{aligned} \tag{6.4.6}$$

that is, the uniform growth term drops out when we consider the changes of the relative frequencies in the population, and the growth rate of p^i depends on the difference of i's own fitness to the average fitness of the population. In particular, even if all the f_i are positive, the relative frequency of type i can still decrease. We now assume that we have some potential function V for (6.4.2), that is,

$$f_i(p) = \frac{\partial V}{\partial p^i} \quad \text{for } i = 1, \ldots, k. \tag{6.4.7}$$

Lemma 6.4.1. *When (6.4.7) holds, then*

$$\dot{p}^i = \sum_j g^{ij} \frac{\partial V}{\partial p^j} \quad \text{for } i = 1, \ldots, k, \tag{6.4.8}$$

where (g^{ij}) is the inverse of the Fisher metric, see (6.3.5) or (6.3.6).

Equation (6.4.8) means that the dynamics for the p^i is a gradient flow w.r.t. the Fisher metric. In particular, $-V(p)$ then is a Lyapunov function, since

$$-\frac{d}{dt}V(p(t)) = \sum_i \frac{\partial V(p)}{\partial p^i}\dot{p}^i(t) = -\sum_{i,j} g^{ij}\frac{\partial V(p)}{\partial p^i}\frac{\partial V(p)}{\partial p^j}$$

which is negative unless $dV(p) = 0$, that is, unless p is a critical of V.

Proof. For every $\xi = (\xi^1, \ldots, \xi^k)$ with $\sum_j \xi^j = 0$, that is, for every tangent vector to the simplex Σ^{k-1}, the space in which the dynamics of p takes place, we have

$$\begin{aligned}
\sum_{i,j} g_{ij}\dot{p}^i\xi^j &= \sum_j \frac{1}{p^j}\dot{p}^j\xi^j \\
&= \sum_j (f_j - \bar{f})\xi^j \\
&= \sum_j f_j\xi^j \\
&= \sum_j \frac{\partial V}{\partial p^j}\xi^j,
\end{aligned}$$

which implies (6.4.8). $\qquad\qquad\qquad\qquad\qquad\qquad\qquad\qquad\qquad\qquad\square$

The condition (6.4.7) is locally equivalent to the integrability condition

$$\frac{\partial f_i}{\partial p^j} = \frac{\partial f_j}{\partial p^i} \text{ for all } i, j. \tag{6.4.9}$$

The more general condition that there exist some function $f_0(p)$ with

$$f_i(p) + f_0(p) = \frac{\partial V}{\partial p^i} \text{ for } i = 1, \ldots, k \tag{6.4.10}$$

is locally equivalent to

$$\frac{\partial f_i}{\partial p^j} + \frac{\partial f_j}{\partial p^l} + \frac{\partial f_l}{\partial p^i} = \frac{\partial f_i}{\partial p^l} + \frac{\partial f_l}{\partial p^j} + \frac{\partial f_j}{\partial p^i} \text{ for all } i, j, l = 1, \ldots, k. \tag{6.4.11}$$

Because a function $f_0(p)$ drops out of the replicator dynamics (6.4.6), (6.4.10) suffices for the conclusion of the Lemma. In fact,

Lemma 6.4.2. *(6.4.11) is necessary and sufficient for Lemma 6.4.1 to hold on a simply connected domain.*

Proof. (See [60]). Let Ω be a simply connected domain in Euclidean space $\mathbb{R}^m$. Then, given two times continuously differentiable functions ϕ_j, $j = 1, \ldots, m$, there exists a function V with

$$\frac{\partial V}{\partial x^j} = \phi_j \text{ for } j = 1, \ldots, m \tag{6.4.12}$$

iff

$$\frac{\partial \phi_j}{\partial x^\ell} = \frac{\partial \phi_\ell}{\partial x^j} \quad \text{for all } j, \ell, \tag{6.4.13}$$

because these are the conditions for the second derivatives of V to commute, $\frac{\partial^2 V}{\partial x^j \partial x^\ell} = \frac{\partial^2 V}{\partial x^\ell \partial x^j}$.

We therefore consider the functions

$$g_j(x^1, \ldots, x^{k-1}) := f_j(x^1, \ldots, x^{k-1}, 1 - x^1 - \cdots - x^{k-1}) \text{ for } j = 1, \ldots, k$$
$$\text{and} \quad \phi_j := g_j - g_k \quad \text{for } j = 1, \ldots, k-1 =: m.$$

If (6.4.11) holds, these functions ϕ_j then satisfy (6.4.13), and so, we can find a function V with (6.4.12).

We then obtain (6.4.10) with $f_0 = g_k$. The proof of Lemma 6.4.1 then yields the claim.
$\qquad\qquad\qquad\qquad\qquad\qquad\qquad\qquad\qquad\qquad\qquad\qquad\qquad\qquad\quad\square$

A special case is the linear case where

$$f_i(p) = \sum_j a_{ij} p^j. \tag{6.4.14}$$

In this case, (6.4.9) becomes the symmetry

$$a_{ij} = a_{ji} \text{ for all } i, j, \tag{6.4.15}$$

and (6.4.11) becomes

$$a_{ij} + a_{jl} + a_{li} = a_{il} + a_{lj} + a_{ji} \text{ for all } i, j, l. \tag{6.4.16}$$

We next include mutations and assume that allele j mutates with probability m_{ij} into i. More precisely, m_{ij} is the mutation rate from j to i, and we assume that in a sufficiently large population, variances of the mutation rate can be neglected so that we obtain the mutation dynamics

$$\dot{p}^i = \sum_j (m_{ij} p^j - m_{ji} p^i). \tag{6.4.17}$$

The combined effects of mutation and selection then are obtained by combining (6.4.6) and (6.4.17):

$$\dot{p}^i = p^i (f_i - \bar{f}) + \sum_j (m_{ij} p^j - m_{ji} p^i). \tag{6.4.18}$$

In order to also represent (6.4.18) as a gradient dynamics, we need to assume that

$$m_{ij} = m_i, \tag{6.4.19}$$

that is, that the mutation rate depends only on the target allele. This is in fact the condition needed to make the analogue of (6.4.11) hold for the right hand side of (6.4.18). We then also define the total mutation rate

$$m := \sum_i m_i. \tag{6.4.20}$$

With

$$F_i := f_i + \frac{m_i}{p^i}, \quad \bar{F} := \sum_i F_i p^i = \bar{f} + m, \tag{6.4.21}$$

Equation (6.4.18) becomes

$$\dot{p}^i = p^i (F_i - \bar{F}). \tag{6.4.22}$$

When (6.4.7), that is, when f is a gradient, then so is F,

$$F_i(p) = \frac{\partial W}{\partial p^i} \tag{6.4.23}$$

with

$$W(p) := V(p) + \sum_i m_i \log p^i \text{ for } i = 1, \ldots, k. \tag{6.4.24}$$

Thus, we have a gradient dynamics again. In any case, (6.4.18) is a *deterministic* equation for the temporal dynamics of the probabilities (relative frequencies) p^i.

This changes when we also include recombination. The basic model of Wright and Fisher for diploid populations is based on the multinomial distribution. As before, we assume that we have a population of size $2N$. One might think of individuals here, but for recombination, one should rather think of gametes. The number of gametes will typically be much larger, perhaps by several orders of magnitude, than the number of individuals in a population. This may serve as a justification for looking at the limit $N \to \infty$.

We assume that the recombination process at the single locus that we consider here, with allele frequencies p_j, can be modelled by sampling with replacement. That is, we assume that in generation t, we have $2p_j N$ alleles of type j, and the probabilities for the numbers N_i of alleles of type i in generation $t + 1$ are given by

$$p(N_1 = n_1, \ldots, N_k = n_k) = \frac{2N!}{n_1! n_2! \ldots n_k!} (p^1)^{n_1} (p^2)^{n_2} \ldots (p^k)^{n_k}. \tag{6.4.25}$$

This is, of course, the generalization of (6.2.1) to the case of more than two alleles. The *expected* relative frequency of alleles of type j in that generation is then

$$p^j, \tag{6.4.26}$$

that is, the same as in the previous generation. The covariance for i, j is

$$p^i(\delta_{ij} - p^j), \tag{6.4.27}$$

see Lemma 3.1.2. (Note that we look at the relative frequencies instead of the absolute ones so that the factor $2N$ drops out of (6.4.26), (6.4.27).) Thus, from (6.4.27), we obtain the inverse Fisher metric (6.3.11).

So far, this is for discrete generation time. In order to pass to continuous time, we need to do a diffusion approximation. This involves a rescaling of time. As in Sect. 6.2, one unit of the new—continuous—time t corresponds now to $2N$ generations or units of discrete time. In general, when we have a Markov chain like (6.4.25), with

$$a_{ij}(p) = p^i(\delta_{ij} - p^j). \tag{6.4.28}$$

$$b_i(p) = \dot{p}^i, \tag{6.4.29}$$

the probability density $\phi(p, q, t)$ that the allele frequencies are p at time t, given that they are q at time 0, is determined by the forward Kolmogorov equation

$$\frac{\partial}{\partial t}\phi(p, q, t) = \frac{1}{2}\sum_{i,j=1}^{k-1}\frac{\partial^2}{\partial p^i \partial p^j}(a_{ij}(p)\phi(p, q, t)) - \sum_{i=1}^{k-1}\frac{\partial}{\partial p^i}(b_i(p)\phi(p, q, t)).$$
$$\tag{6.4.30}$$

This generalizes (6.2.12). For the derivation of Kolmogorov equations within the context of dynamical systems, we refer to [87], within the context of Markov chains to [41]. In [62], we shall derive (6.4.30) in a direct manner within the framework established here.

Equation (6.4.30) now is a *deterministic* equation for the probability density ϕ for having the relative allele frequencies p^i. The first order term on the right hand side is called a drift term; it represents the dynamics of the expectation value. The second order term is called a diffusion term. This may become a little confusing because (6.4.25) described random genetic drift. Thus, in the diffusion approximation, genetic drift causes a diffusion and not a drift term.

Bibliography

1. Alberts, B., Johnson, A., Walter, P., Lewis, J., Raff, M., Roberts, K.: Molecular Biology of the Cell, vol. 5. Garland Science, New York (2007)
2. Altschul, S., Gish, W., Miller, W., Myers, E., Lipman, D.: Basic local alignment search tool. J. Mol. Bio. **215**, 403–410 (1990)
3. Atay, F. (ed.): Complex Time-Delay Systems: Theory and Applications. Springer, Berlin (2010)
4. Atay, F., Jost, J.: Qualitative inference in dynamical systems. In: Stumpf, M., Balding, D., Girolami, M. (eds.) Handbook of Statistical Systems Biology. J. Wiley, Chichester (2011)
5. Ay, N., Jost, J., Lê, H.V., Schwachhöfer, L.: Information geometry, in preparation
6. Banerjee, A., Jost, J.: Laplacian spectrum and protein-protein interaction networks, arXiv, 0705.3373 (2007)
7. Banerjee, A., Jost, J.: On the spectrum of the normalized graph Laplacian. Lin. Alg. Appl. **428**, 3015–3022 (2008)
8. Banerjee, A., Jost, J.: Graph spectra as a systematic tool in computational biology. Discrete Appl. Math. **157**, 2425–2431 (2009)
9. Banerjee, A., Jost, J.: Spectral plots and the representation and interpretation of biological data. Theory Biosc. **126**, 15–21 (2007)
10. Banerjee, A., Jost, J.: Spectral plot properties: towards a qualitative classification of networks. Netw. Heterogen. Media **3**, 395–411 (2008)
11. Banerjee, A., Bauer, F., Jost, J.: Eigenvalues of the normalized graph Laplacian and the architecture of directed networks, to appear (2012)
12. Bandelt, H.-J., Dress, A.: A canonical decomposition theory for metrics on a finite set. Adv. Math. **92**, 47–105 (1992)
13. Barash, Y., Calarco, J.A., Gao, W., Pan, Q., Wang, X., Shai, O., Blencowe, B.J., Frey, B.J.: Deciphering the splicing code. Nature **465**, 53–59 (2010)
14. Bauer, F.: Normalized graph Laplacians for directed graphs. Lin. Alg. Appl. **436**, 4193–4222 (2012)
15. Bauer, F., Jost, J.: Bipartite and neighborhood graphs and the spectrum of the normalized graph Laplacian. Comm. Anal. Geom. **436**, 4193–4222 (2012)
16. Bender, E., Canfield, E.: The asymptotic number of labeled graphs with given degree sequences. J. Comb.Th. (A). **24**, 296–307 (1978)
17. Berglund, Gentz, N.B.: Noise-Induced Phenomena in Slow-Fast Dynamical Systems. Springer, New York (2005)
18. Bolobás, B.: Random Graphs. Cambridge University Press, Cambridge (2001)

J. Jost, *Mathematical Methods in Biology and Neurobiology*, Universitext,
DOI: 10.1007/978-1-4471-6353-4, © Springer-Verlag London 2014

19. Bolobás, B.: Modern Graph Theory. Springer, New York (1998)
20. Breidbach, O., Jost, J.: Working in a multitude of trends: species balancing populations. J. Zool. Res. Evol. Syst. **42**, 202–207 (2004)
21. Brown, J., West, G. (eds.): Scaling in Biology. Oxford University Press, Oxford (2000)
22. Bürger, R.: The Mathematical Theory of Selection, Recombination, and Mutation. Wiley, New York (2000)
23. Butazzo, G., Giaquinta, M., Hildebrandt, S.: One-Dimensional Variational Problems. Oxford University Press, Oxford (1998)
24. Carroll, S.: Endless Dreams Most Beautiful. W.W Norton, New York (2005)
25. Chavel, I.: Isoperimetric Inequalities. Cambridge University Press, Cambridge (2001)
26. Chen, K., Rajewsky, N.: The evolution of gene regulation by transcription factors and microRNAs—review. Nat. Rev. Genet. **8**, 93–103 (2007)
27. Chueh, K.N., Conley, C., Smoller, J.: Positively invariant regions for systems of nonlinear diffusion equations. Indiana Univ. Math. J. **26**, 373–392 (1977)
28. Chung, F.: Spectral Graph Theory. American Mathematical Society, New York (1997)
29. Colonius, H., Schulze, H.: Tree structures for proximity data. British J. Math. Statist. Psychol. **34**, 167–180 (1981)
30. Courant, R., Friedrichs, K., Lewy, H.: Über die partiellen Differentialgleichungen der mathematischen Physik. Math. Ann. **100**, 32–74 (1928)
31. dal Maso, G.: An Introduction to Γ-Convergence. Birkhäuser, Basel, Boston (1993)
32. Darwin, C.: The Origin of Species. John Murray, London (1859) [Reprint: Penguin Books, London (1968)]
33. Dayan, P., Abbott, L.F.: Theoretical Neuroscience. MIT Press, Cambridge (2001)
34. Djebali, S., et al.: Landscape of transcription in human cells. Nature **489**, 101–108 (2012)
35. Dodziuk, J.: Difference equations, isoperimetric inequality and transience of certain random walks. Trans. Amer. Math. Soc. **284**, 787–797 (1984)
36. Dress, A.: Recent results and new problems in phylogenetic combinatorics (2002)
37. Dress, A., Huber, K., Koolen, J., Moulton, V., Spillner, A.: Basic Phylogenetic Combinatorics. Cambridge University Press, Cambridge (2012)
38. ENCODE, Project Consortium: Identification and analysis of functional elements in 1 % of the human genome by the ENCODE pilot project. Nature **447**, 779–796 (2007)
39. Erdös, P., Rényi, A.: On random graphs, I. Publ. Math. Debrecen **6**, 290–291 (1959)
40. Ermentrout, G.B., Terman, D.: Mathematical Foundations of Neuroscience. Springer, New York (2010)
41. Ethier, S.N., Kurtz, T.G.: Markov Processes. Characterization and Convergence. Wiley Series in Probability and Mathematical Statistics. Wiley, New York (1986)
42. Evans, L.C.: Partial Differential Equations. American Mathematical Society, 2nd edn. New York (2010)
43. Ewens, W.: Mathematical Population Genetics I. Theoretical Introduction. Springer, Heidelberg (2004)
44. Fenichel, N.: Geometric singular perturbation theory for ordinary differential equations. J. Diff. Equat. **31**, 53–98 (1979)
45. Fisher, R.A.: The Genetical Theory of Natural Selection. Clarendon Press, Oxford (1930)
46. Futuyma, D.: Evolutionary Biology. Sinauer, Massachusetts (1997)
47. Gavrilets, S.: Fitness Landscapes and the Origin of Species. Princeton University Press, Princeton (2004)
48. Gladwell, G., Davies, E., Leydold, J., Stadler, P.: Discrete nodal domain theorems. Lin. Alg. Appl. **336**, 51–60 (2001)
49. Gencode, Version 14. http://www.gencodegenes.org/releases/14.html. Accessed 21 June 2012
50. Godsil, C., Royle, G.: Algebraic Graph Theory. Springer, New York (2001)
51. Goldberg, D.: Genetic Algorithms. Dorling Kindersley, New Delhi (2008)
52. Goldbeter, A.: Biochemical Oscillations and Biochemical Rhythms: The Molecular Bases of Periodic and Chaotic Behaviour. Cambridge University Press, Cambridge (1996)

53. Goldbeter, A., Lefever, R.: Dissipative structures for an allosteric model. Applications to glycolytic oscillations. Biophys. J. **12**, 1302–1315 (1972)
54. Green, R., et al.: A draft sequence of the Neanderthal genome. Science **328**, 710–722 (2010)
55. Grimmett, G., Stirzacker, D.: Probability and Random Processes. Oxford University Press, Oxford (2001)
56. Haccou, P., Jagers, P., Vatutin, V.A.: Branching Processes. Cambridge University Press, Cambridge (2005)
57. Hein, J., Schierup, M., Wiuf, C.: Gene Genealogies, Variation and Evolution. Oxford University Press, Oxford (2005)
58. Hennig, W.: Phylogenetic Systematics. University Illinois Press, Champaign (1966)
59. Hofacker, I.L., Fontana, W., Stadler, P.F., Bonhoeffer, L.S., Tacker, M., Schuster, P.: Fast folding and comparison of RNA secondary structures. Monatsh. Chemie **125**, 167–188 (1994)
60. Hofbauer, J., Sigmund, K.: The Theory of Evolution and Dynamical Systems. Cambridge University Press, Cambridge (1988)
61. Hofbauer, J., Sigmund, K.: Evolutionary Games and Population Dynamics. Cambridge University Press, Cambridge (1998)
62. Hofrichter, J., Jost, J., Tran, T.D.: Information geometry and population genetics, in preparation
63. Nature Initial sequencing and analysis of the human genome. **409**, 860–921 (2001)
64. Izhikevich, E.: Dynamical Systems in Neuroscience. MIT Press, Cambridge (2007)
65. Jonsson, J.: Simplicial Complexes of Graphs. Springer LNM 1928, New York (2008)
66. Jost, J.: On the Notion of Fitness, or: The Selfish Ancestor. Theory Biosci. **121**, 331–350 (2003)
67. Jost, J.: Postmodern Analysis. 3rd edn. Springer, Berlin (2005)
68. Jost, J.: Partial Differential Equations. 3rd edn. Springer, Berlin (2013)
69. Jost, J.: Dynamical Systems. Springer, Berlin (2005)
70. Jost, J.: Riemannian Geometry and Geometric Analysis. 6th edn. Springer, Berlin (2011)
71. Jost, J.: Mathematical neurobiology: Concepts, tools, and questions, to appear
72. Jost, J.: Biologie und Mathematik, to appear
73. Jost, J.: Mathematical Concepts, to appear
74. Jost, J., X. Li-Jost. Calculus of variations. Cambridge University Press, Cambridge (1998)
75. Jost, J., Pepper, J.: Individual optimization efforts and population dynamics: a mathematical model for the evolution of resource allocation strategies, with applications to reproductive and mating systems. Theory Biosc. **127**, 31–43 (2008)
76. Jost, J., Scherrer, K.: Information theory, gene expression, and combinatorial regulation - A quantitative analysis, to appear in Theory Biosc
77. Kahle, T., Bertschinger, N., Ay, N., Jost, J., Olbrich, E.: Quantifying structure in networks. Europ. Phys. J. B **77**, 239–247 (2010)
78. Keller, E., Segel, L.: Initiation of slime mold aggregation viewed as an instability. J. Theor. Biol. **26**, 399–415 (1970)
79. Keller, E., Segel, L.: Model for chemotaxis. J. Theor. Biol. **30**, 225–234 (1971)
80. Kimmel, M., D.: Axelrod, Branching processes in biology. Springer, New York (2002)
81. Kingman, J.: The coalescent. Stoch. Proc. Appl. **13**, 235–248 (1982)
82. Klipp, E., Herwig, R., Kowald, A., Wierling, C., H.: Lehrach, Systems biology in practice. Wiley, Weinheim (2005)
83. Koch, C.: Biophysics of computation. Oxford University Press, New York (1999)
84. Kolmogoroff, A., Petrovsky, I., Piscounoff, N.: Étude de l' équation de la diffusion avec croissance de la quantité de la matière et son application à un problème biologique. Moscow Univ. Bull. Math. **1**, 1–25 (1937)
85. Kondrashov, A.S.: Classification of hypotheses on the advantage of amphimixis. J. Hered. **84**, 372–387 (1993)
86. Kosiuk, I., Smolyan, P.: Scaling in singular perturbation problems: Blowing up a relaxation oscillator. SIAM J. Appl. Dyn. Sys. **10**, 1307–1343 (2011)
87. Lasota, A., Mackey, M.: Chaos, Fractals, and Noise. Springer, Berlin (1994)

88. Margulis, L.: Symbiotic Planet: A New Look At Evolution, Basic Books (1999)
89. Meinhardt, H., de Boer, P.: Pattern formation in Escherichia coli: A model for the pole-to-pole oscillations of Min proteins and the localization of the division site. PNAS **98**, 14202–14207 (2001)
90. Mendoza, M. et al.: An African American parental lineage adds an extremely ancient root to the human Y chromosome phylogenetic tree. Am er.J.Hum.Gen. 92, 454–459, 2013
91. Mitchell, M.: An Introduction to Genetic Algorithms. MIT Press, Cambridge (1998)
92. Murray, J.: Mathematical Biology, 2 vols. 3rd edn. Springer, Heidelberg (2008)
93. Nelson, D., Cox, M.: Lehninger: Principles of Biochemistry. W.H. Freeman, New York (2005)
94. Newman, M.: Random graphs as models of networks. In: Bornholdt, S., Schuster, H.G. (eds.) Handbook of Graphs and Networks. Wiley, Berlin (2002)
95. Newman, M.: Networks. Oxford University Press, Oxford (2010)
96. Øksendal, B.: Stochastic Differential Equations. Springer, Berlin (2008)
97. Pavliotis, G., Stuart, A.: Multiscale methods. Averaging and Homogenization. Springer, Berlin (2008)
98. Peña, B., Pérez-García, C.: Stability of Turing patterns in the Brusselator model. Phys. Rev. E **64**, 056213 (2001)
99. Perthame, B.: Transport Equations in Biology. Birkhäuser, Boston (2007)
100. Phillipson, P., Schuster, P.: An analytical picture of neuron oscillations. Int. J. Bifurc. Chaos **14**, 1539–1548 (2004)
101. Quiring, R., Walldorf, U., Kloter, U., Gehring, W.: Homology of the eyeless gene of Drosophila to the Small eye gene in mice and Aniridia in humans. Science **265**, 785–789 (1994)
102. Rice, S.: Evolutionary Theory. Sinauer, Massachusetts (2004)
103. Scherrer, K., Jost, J.: The gene and the genon concept : A functional and information-theoretic analysis. Mol. Syst. Biol. **3**, 87 (2007)
104. Scherrer, K., Jost, J.: Gene and genon concept: Coding vs. regulation. Theory in Biosciences **126**, 65–113 (2007)
105. Scherrer, K., Jost, J.: Response to commentaries on our paper Gene and genon concept: Coding vs. regulation. Theory in Biosciences **128**, 171–177 (2009)
106. Schindler, S., Breidbach, O., Jost, J.: Preferring the fittest mates: An analytically tractable model. J. Theo. Biol. **317**, 30–38 (2013)
107. Semple, C., Steel, M.: Phylogenetics. Oxford University Press, Oxford (2003)
108. Shepherd, G.: Neurobiology. Oxford University Press, Oxford (1994)
109. Shuster, S., Wade, M.: Mating systems and strategies. Princeton University Press, New Jersey (2003)
110. Smoller, J.: Shock Waves and Reaction-Diffusion Equations. 2nd edn. Springer, New York (1994)
111. Sporns, O.: Networks of the Brain. MIT Press, Cambridge (2011)
112. Steel, M., Linz, S., Huson, D., Sanderson, M.: Identifying a species tree subject to random lateral gene transfer. J. Theo. Biol. **322**, 81–93 (2013)
113. Stevens, A., Othmer, H.: Aggregation, blow-up, and collapse: The ABC's of taxis in reinforced random walks. SIAM J. Appl. Math. **57**, 1044–1081 (1997)
114. Strimmer, K., von Haeseler, A.: Quartet puzzling: a quartet maximum likelihood method for reconstructing tree topologies. Mol. BiolEvol. **13**, 964–996 (1996)
115. Tuckwell, H.: Stochastic Processes in the Neurosciences. SIAM, Philadelphia (1988)
116. Tuckwell, H.: Introduction to Theoretical Neurobiology, vol. 2. Cambridge University Press, Cambridge (1988)
117. Turing, A.: The chemical basis of morphogenesis. Phil. Trans. Roy. Soc. **B327**, 37–72 (1952)
118. Van Dam, E., Haemers, W.: Which graphs are determined by their spectrum?. Lin. Alg. Appl. 373, 241–272, 2003
119. Vincent, A., Goldenberg, S., Standart, N., Civelli, O., Imaizumi-Scherrer, M.T., Maundrell, K., Scherrer, K.: Potential role of mRNP proteins in cytoplasmic control of gene expression in duck erythroblasts. Mol. Biol. Rep. **7**, 71–81 (1981)
120. Walgraef, D.: Spatio-temporal pattern formation. Springer, New York (1997)
121. Zeidler, E.: Nonlinear Functional Analysis and its Applications, vol. 3. Variational Methods and Optimization. Springer, New York (1984)

Index

J. Jost, *Mathematical Methods in Biology and Neurobiology*, Universitext,
DOI: 10.1007/978-1-4471-6353-4, © Springer-Verlag London 2014